Math Bridge

7th grade

Written by:

Tracy Dankberg & James Michael Orr

Project Directors: Michele D. Van Leeuwen
Scott G. Van Leeuwen

Creative & Marketing Director: George Starks

Design & Technical Project Director: Dante J. Orazzi

TABLE OF CONTENTS

Woods

Introduction

The *Math Bridge* series is designed to help students improve their mathematical skills in all areas. This book has been developed to provide seventh grade students with skill-based exercises in the following areas: whole numbers; decimals; number theory; fractions; ratio and proportion; percent; geometry; measurement; integers; pre-algebra; probability and statistics. The purpose of this book, therefore, is to strengthen students' mathematical concepts, thus helping them to become better mathematicians and to improve achievement test scores.

Math Bridge includes many extras to help your students in their study of mathematics. For instance,

- ✔ An Incentive Contract begins the book to motivate students to complete their work.
- ✔ A diagnostic test has been included to help assess your students' mathematical knowledge.
- ✔ Exercises become progressively more difficult as students work through the book.
- ✔ Tips are included throughout the book as reminders to help students successfully complete their work.
- ✔ Thought-provoking questions (Think About It) are periodically placed throughout the book to emphasize critical thinking skills.
- ✔ Additional exercises are included to help students in practicing with estimation.
- ✔ The exercises prepare students for standardized achievement tests.
- ✔ Each section includes problem-solving exercises written with the purpose of reinforcing the skills taught in that section.

Mathematics is all around us and is an essential part of life. It is the authors' intention that through the completion of this book, students will come away with a stronger knowledge of mathematics to assist them both inside and outside of the classroom.

Incentive Contract

In • cen'tive, ***n.*** **1.** Something that urges a person on. **2.** Enticing. **3.** Encouraging. **4.** That which excites to action or moves the mind.

LIST YOUR AGREED-UPON INCENTIVE FOR EACH SECTION BELOW

Place a ✔ after each activity upon completion

Student's Signature ______________________________

Parent or Teacher Signature ____________________

PG	Activity Title	✔
9	Estimating	
10	Adding & Subtracting	
11	Multiplying & Dividing	
12	Powers & Exponents	
13	Order of Operations	
14	Problem Solving	

MY INCENTIVE IS ✖

PG	Activity Title	✔
15	Comparing	
16	Ordering	
17	Estimating Addition & Subtraction	
18	Adding & Subtracting	
19	Estimating Multiplication & Division	
20	Multiplying & Dividing	
21	Scientific Notation	
22	Problem Solving	

MY INCENTIVE IS ✖

PG	Activity Title	✔
23	Writing & Evaluating Algebraic Expressions	
24	Writing Algebraic Equations	
25	Problem Solving	
26	Solving Addition & Subtraction Equations	
27	Solving Multiplication & Division Equations	
28	Solving More Equations	
29	Solving Two-Step Equations	
30	Problem Solving	

MY INCENTIVE IS ✖

PG	Activity Title	✔
31	Meaning of Integers	
32	Absolute Value	
33	Comparing & Ordering	
34	Multiplying & Dividing	
35	Adding Integers	
36	Subtracting Integers	
37	Solving Equations	
38	Problem Solving	

MY INCENTIVE IS ✖

PG	Activity Title	✔
39	Reading Charts & Tables	
40	Graphing Skills	
41	Predicting Outcomes	
42	Mean, Median & Mode	
43	Problem Solving	

MY INCENTIVE IS ✖

PG	Activity Title	✔
44	Classifying Angles	
45	Classifying Triangles	
46	Classifying Quadrilaterals	
47	Angle Sums	
48	Symmetry	
49	Coordinate Planes	
50	Prime Factorization	
51	Greatest Common Factor	
52	Lowest Common Multiple	
53	Fractions as Decimals, Decimals as Fractions	
54	Comparing & Ordering	
55	Simplifying Fractions	
56	Improper Fractions & Mixed Numbers	
57	Problem Solving	
58	Adding & Subtracting	
59	Adding & Subtracting Mixed Numbers	

MY INCENTIVE IS ✖

PG	Activity Title	✔
60	Multiplying Fractions & Mixed Numbers	
61	Dividing Fractions & Mixed Numbers	
62	Solving Equations	
63	Problem Solving	

MY INCENTIVE IS ✖

PG	Activity Title	✔
64	Customary System	
65	The Metric System	
66	Perimeter & Area of Quadrilaterals	
67	Squares & Square Roots	
68	Pythagorean Theorem	
69	Area of Triangles & Trapezoids	
70	Circumference & Area of Circles	
71	Problem Solving	

MY INCENTIVE IS ✖

PG	Activity Title	✔
72	Ratios & Equal Ratios	
73	Rates	
74	Solving Proportions	
75	Problem Solving	

MY INCENTIVE IS ✖

PG	Activity Title	✔
76	What is a Percent?	
77	Percents & Fractions	
78	Percents & Decimals	
79	Large & Small Percents	
80	Percent of a Number	
81	Percent of Change	
82	Discounts & Sales Tax	
83	Problem Solving	

MY INCENTIVE IS ✖

PG	Activity Title	✔
84	Arithmetic & Geometric Sequences	
85	Problem Solving	

MY INCENTIVE IS ✖

PG	Activity Title	✔
86	Tree Diagrams	
87	Fundamental Counting Principle	
88	Experimental & Theoretical Probability	
89	Problem Solving	

MY INCENTIVE IS ✖

DIAGNOSTIC TEST

Name ____________________ **Score** __________

Directions: Read the following problems. For each question, fill in the circle of the correct answer. If the correct answer is not given, fill in the answer space marked **N** (Not Given).

1. $832 - 749 =$ O A. 93 O B. 83 O C. 193 O D. 183

2. $2383 + 298 =$ O A. 2581 O B. 2680 O C. 2682 O D. N

3. $674 \div 37 =$ O A. 18.21 O B. 18.29 O C. 18.22 O D. 18.26

4. $345 \times 31 =$ O A. 9,365 O B. 10,695 O C. 9,695 O D. 10,035

5. $12 + (3 \times 13) =$ O A. 61 O B. 51 O C. 195 O D. N

6. $\frac{9 + (36 \div 6)}{3} =$ O A. 11 O B. 5 O C. 9 O D. 15

7. $31.02 - .85 =$ O A. 30.27 O B. 30.17 O C. 31.17 O D. N

8. $73.63 - 1.09 =$ O A. 72.64 O B. 72.54 O C. 71.64 O D. 71.54

9. $6.84 + 3.982 =$ O A. 4.666 O B. 46.66 O C. 10.822 O D. 108.22

10. $638 \times 0.06 =$ O A. 3.828 O B. 38.28 O C. .3828 O D. 382.8

11. $72.8 \times 4.71 =$ O A. 3428.888 O B. 34.288 O C. 342.888 O D. N

12. $538 \div 38 =$ O A. 1.415 O B. 14.16 O C. 141.51 O D. 14.13

13. $63.07 \div 7.42 =$ O A. .85 O B. 8.5 O C. 85 O D. 850

14. What is the closest estimate of 2131×6?
    O A. 22,000 O B. 300 O C. 12,000 O D. 1,200

15. The closest estimate of 297×200 is what?
    O A. 49,000 O B. 50,000 O C. 59,000 O D. 60,000

16. What is the closest estimate of $73.3 \div 9.25$?
    O A. 7 O B. 8 O C. 9 O D. 10

17. The closest estimate of \$61.93 – \$49.96 is what?
    O A. \$2.00 O B. \$10.00 O C. \$11.00 O D. \$12.00

18. What is the closest estimate of \$18.95 + \$6.27?
    O A. \$24 O B. \$25 O C. \$27 O D. \$28

DIAGNOSTIC TEST

Name ______________________

19. What is another way to write 5^4?

O A. $5 + 5 + 5 + 5$ O B. $5 + 4$ O C. $5 \times 5 \times 5 \times 5$ O D. 5×4

20. What is the greatest common factor of 18, 54, and 90?

O A. 3 O B. 6 O C. 9 O D. 18

21. What is the least common multiple of 8 and 12?

O A. 4 O B. 96 O C. 24 O D. 32

22. Which of the following numbers is prime?

O A. 2 O B. 15 O C. 21 O D. 51

23. Which of the following numerals is the same as $\frac{16}{24}$?

O A. $\frac{1}{2}$ O B. $\frac{3}{4}$ O C. $\frac{2}{3}$ O D. $\frac{5}{8}$

24. Which one of the following decimals is equal to $\frac{1}{8}$?

O A. .0125 O B. .125 O C. 1.25 O D. 12.5

25. What is the reciprocal of $\frac{5}{6}$?

O A. $\frac{6}{5}$ O B. $\frac{5}{6}$ O C. 2 O D. 0

26. What would you replace the x with in the following equation? $\frac{12}{x} \times \frac{3}{6} = \frac{36}{78}$

O A. 12 O B. 9 O C. 14 O D. 13

27. Look at the following fractions in the box. How many are greater than $\frac{5}{10}$?

$$\frac{5}{12}, \frac{7}{16}, \frac{2}{3}, \frac{6}{12}, \frac{3}{4}, \frac{3}{8}, \frac{5}{8}$$

O A. 3 O B. 4 O C. 5 O D. 6

28. What is $6\frac{3}{4}$ written as an improper fraction?

O A. $\frac{24}{4}$ O B. $\frac{24}{6}$ O C. $\frac{27}{4}$ O D. $\frac{27}{6}$

29. $1\frac{1}{2} - \frac{7}{8} =$

O A. $\frac{5}{8}$ O B. $\frac{3}{4}$ O C. $1\frac{1}{4}$ O D. $\frac{1}{4}$

DIAGNOSTIC TEST

Name ______________________

30. $\frac{12}{14} + \frac{5}{7} =$

O A. $\frac{23}{14}$ O B. $1\frac{3}{14}$ O C. $1\frac{4}{7}$ O D. $1\frac{5}{7}$

31. $8 \times \frac{3}{4} =$

O A. 4 O B. 5 O C. 6 O D. 7

32. $\frac{3}{8} \times \frac{2}{3} \times \frac{1}{2} =$

O A. $\frac{7}{48}$ O B. $\frac{1}{8}$ O C. $\frac{6}{13}$ O D. $\frac{1}{4}$

33. $\frac{3}{5} \div \frac{4}{10} =$

O A. $1\frac{1}{2}$ O B. $1\frac{2}{5}$ O C. $1\frac{9}{20}$ O D. $\frac{1}{5}$

34. $24 \div \frac{2}{4} =$

O A. 12 O B. 24 O C. 48 O D. 60

35. Which of the following pairs of fractions forms a proportion?

O A. $\frac{6}{14}, \frac{3}{8}$ O B. $\frac{5}{7}, \frac{4}{6}$ O C. $\frac{6}{9}, \frac{8}{12}$ O D. $\frac{5}{9}, \frac{18}{27}$

36. In Veronica's neighborhood, the ratio of cats to dogs is 3 to 4. If there are 28 dogs, how many cats are there?

O A. 14 O B. 21 O C. 7 O D. 18

37. Which of the following percents is equal to $\frac{7}{10}$?

O A. .7% O B. 7% O C. 70% O D. 700%

38. Which of the following decimals is equal to 23%?

O A. .023 O B. .23 O C. 2.3 O D. 23

39. What is 20% of 40?

O A. 8 O B. 10 O C. 18 O D. 20

40. The French Club had a fundraising campaign during the month of November, making a profit of $425. The club collected $1,375 in sales. About what percentage of the total amount collected was profit?

O A. 3% O B. 30% O C. 36% O D. 300%

DIAGNOSTIC TEST

Name ______________________________

41. In the fall book sale, Harriet ordered a three-volume set of books for $31.25, instead of paying $12.95 for each of the three books. By buying the set, about what percent did she save?

   O A. 30% O B. 80% O C. 20% O D. 66%

42. What is the perimeter of the rectangle? 3 in, 7 in

   O A. 10 in O B. 20 in O C. 21 in O D. 17 in

43. What is the area of the triangle? 4 cm, 7 cm, 6 cm

   O A. 17 cm^2 O B. 12 cm^2 O C. 24 cm^2 O D. 21 cm^2

44. What is the circumference of the circle? (Hint: $C = \Pi d$)

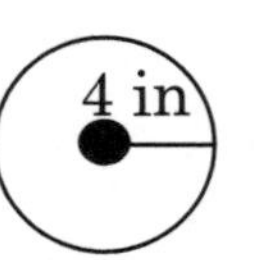

   O A. 6.28 in O B. 12.56 in O C. 25.12 in O D. 50.24 in

45. If there are 36 eggs in 3 dozen, approximately how many eggs are there in 9 dozen?

   O A. 72 O B. 95 O C. 110 O D. 125

46. Camille has a swimming pool that holds 30 kiloliters of water. If it takes her 5 minutes to add 10 kiloliters to the pool, how many minutes would it take to fill a 35 kiloliter pool?

   O A. $12\frac{1}{2}$ O B. 15 O C. $17\frac{1}{2}$ O D. 20

47. If one wanted to subtract 6 lb 8 oz from 8 lb 5 oz, one would first convert 8 lb 5 oz to what?

   O A. 8 lb 13 oz O B. 7 lb 5 oz O C. 7 lb 21 oz O D. 8 lb 16 oz

48. Don has 48 black chips and 64 red chips in a bucket. If he chooses one chip from the bucket, the odds are what in favor of choosing a black one?

   O A. 4 to 3 O B. 3 to 7 O C. 1 to 8 O D. 1 to 2

49. What is the mean of the following numbers? 8, 12, 4, 6, 3, 14, 2

   O A. 11 O B. 6 O C. 7 O D. 8

50. Which numeral below makes the number sentence *true*? $\mathbf{x < {}^{-}9}$

   O A. –8 O B. 8 O C. –10 O D. 11

51. $^{-}60 \div 12 =$ O A. –5 O B. 5 O C. 6 O D. –6

52. $7 + {}^{-}4 =$ O A. 11 O B. 3 O C. $^{-}3$ O D. $^{-}11$

53. What is the solution to $y \div 32 = 16$?

   O A. 2 O B. 512 O C. 16 O D. 48

Whole Numbers: Estimating

Name ______________________

Estimate each answer.

Strategy 1: Use compatible numbers.

$3,300 \div 8 =$ 3,300 is close to 3,200, and $3,200 \div 8 = 400$; therefore, a good estimate would be **400.**

---

1. $198 \times 5 =$
2. $79 \div 11 =$
3. $424 \div 6 =$
4. $53 \times 78 =$
5. $213 \div 7 =$
6. $1,718 \div 9 =$
7. $63 \times 59 =$
8. $2,715 \div 92 =$
9. $5,018 \div 2,497 =$
10. $249 \div 6 =$
11. $375 \times 12 =$
12. $353 \div 72 =$

Strategy 2: Use rounding. Round numbers to the highest place value.

$22 \times 96 =$
↓ ↓
$20 \times 100 =$ **2000**

$588 - 413 =$
↓ ↓
$600 - 400 =$ **200**

13. $72 + 11 =$
14. $19.99 + 29.99 =$
15. $19 + 780 =$
16. $98 - 21 =$
17. $381 - 102 =$
18. $525 - 224 =$
19. $79 \times 11 =$
20. $4,801 + 4507 =$
21. $4,839 + 3,837 =$
22. $61 + 41 =$
23. $3,899 + 3,480 =$
24. $16 \times 101 =$

Whole Numbers: Adding & Subtracting

Name____________________

Find each sum or difference.

5,968 + 6,754

```
  5,968
+ 6,754
 12,722
```

1. Line up the digits from the ones place.
2. Add or subtract from right to left.
3. Carry (addition) or borrow (subtraction) as necessary.

---

1. 14,543 + 21,772
2. 56,748 − 5,123
3. 5,261 + 38,239
4. 53,866 − 3,968
5. 65,426 + 22,895
6. 74,231 − 39,132
7. 47,562 + 82,685
8. 20,007 − 14,258
9. 65,854 + 44,338
10. 59,647 − 25,958
11. 57,521 − 6,534
12. 5,724 + 31,085
13. 66,995 − 27,836
14. 74,847 − 3,928
15. 20,020 − 19,985
16. 89,322 − 27,879
17. 65,432 + 26,358 + 72,343 + 52,345
18. 4,235 + 25,876 + 34,526 + 47,878
19. 43,224 + 74,132 + 22,586 + 53,490
20. 52,350 + 67,335 + 76,784 + 32,287
21. 19,500 − 8,123 =
22. 4,152 + 16,894 =
23. 18,959 + 672 =

Whole Numbers: Multiplying & Dividing Name ____________

Find each product or quotient. Study the examples below.

MATH FACTS

Mathematics is a science that helps build our society through construction, engineering, architecture, chemistry, biology & physics!

167 × 54

```
   167
 ×  54
   668
+ 8,350
 9,018
```

808 ÷ 52

```
   15 R 28
52|808
  - 52
    288
  - 260
     28
```

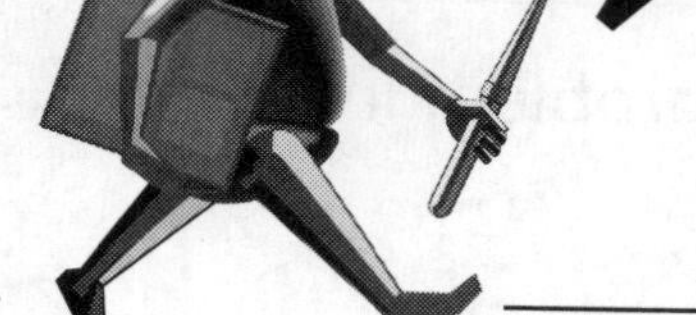

1. 785 × 60 =

2. 5,764 × 500 =

3. 68,834 ÷ 7 =

4. 8,236 × 58 =

5. 4,562 ÷ 34 =

6. 675 × 574 =

7. 35 $\overline{)2,425}$

8. 2,984 × 327 =

9. 67,452 ÷ 73 =

10. 5,067 × 228 =

11. 13,675 ÷ 17 =

12. 841 $\overline{)9,957}$

Whole Numbers: Powers & Exponents

Name ____________________

$4 \cdot 4 \cdot 4 \cdot 4 \cdot 4 = 4^5$ (5 → *exponent*; 4 → *base*)

An ***exponent*** indicates the number of times the ***base*** is used as a factor.

$6^3 = \underline{6 \cdot 6 \cdot 6}$	Expanded form.
$2 \cdot 2 \cdot 3 \cdot 2 = \underline{2^3} \cdot 3$	Exponent form.
$7^2 = \underline{49}$	Simplified.

Write each problem in expanded form.

1. 7^5
2. 2^6
3. 3^5
4. 5^3
5. 4^4
6. 12^2
7. 10^5
8. 5^4
9. 4^3

Write in exponent form.

10. $2 \cdot 2 \cdot 2$
11. $13 \cdot 13$
12. $6 \cdot 5 \cdot 6 \cdot 5$
13. $5 \cdot 5 \cdot 4$
14. $7 \cdot 8 \cdot 7 \cdot 8 \cdot 7$
15. $8 \cdot 9 \cdot 8 \cdot 4 \cdot 4$
16. $2 \cdot 4 \cdot 4 \cdot 2 \cdot 4$
17. $9 \cdot 9 \cdot 10$
18. $10 \cdot 11 \cdot 10 \cdot 11$

Simplify.

19. 11^2
20. $5^3 \cdot 10^2$
21. $7^4 + 3^2$
22. $3^3 \cdot 2^4 \cdot 4^2$
23. $10^2 + 6^3$
24. $9^5 \cdot 6 \cdot 2^2$

Whole Numbers: Order of Operations

Name ____________________

Evaluate each expression. Remember to use the correct order of operations.

MATH FACTS

The Egyptians used a version of modern mathematics, called Hieroglyphic Numeration, to build the Great Pyramids and to chart the stars!

Order of operations:

1. Work inside the parentheses first.
2. Evaluate exponents.
3. Next, multiply and divide, in order *from left to right.*
4. Last, add and subtract, in order *from left to right.*

$(13 - \underline{3^2}) \times 4$

$(\underline{13 - 9}) \times 4$

$4 \times 4 = 16$

---

1. $(24 - 6) \times 2$
2. $24 \div 6 \times 2$
3. $24 \div (6 \times 2)$
4. $24 \times 6 - 2$
5. $24 - 6 \times 2$
6. $24 - 6 + 2$
7. $(5 + 2) \times 4 + 3$
8. $5 + 2 \times (4 + 3)$
9. $(5 + 2) \times (4 + 3)$
10. $5 + 2 \times 4 + 3$
11. $(5^2 - 5) \times 5 \div 25$
12. $(27 \div 3^2) \times 7$
13. $9 + 4^2 - 2$
14. $17 - (10 - 7)$
15. $(17 - 10) - 7$

THINK ABOUT IT!

16. Use any combination of addition, subtraction, multiplication, or division signs in the blanks to make the statement(s) true.

15 ___ 3 ___ 17 ___ 11 = 205

Whole Numbers: Problem Solving

Name ______________________

Solve each problem.

1. Stacy collects money from her newspaper route three days a week. Each day she collects from 21 families. How many families does she collect from each week?

2. If Stacy collects $5 a week from each family, how much does she collect in one week? in one year?

3. Charlie runs a lemonade stand on Saturday afternoons for three hours. He averages 25 customers per hour. If he charges $1 for a tall glass of lemonade, how much money will he make?

4. The school band played 12 pieces of music during their two-hour performance. What was the average time in minutes of each piece?

5. Jenny bought 4 tickets to the school dance. Each ticket cost $3. How much did she pay in all? If she paid with a $20 bill, how much change did she get back?

6. Suppose the 212 people on an airplane each had 2 pieces of luggage weighing an average of 45 pounds each. Estimate how much these bags weigh altogether.

7. Walter bought 4 tickets to a concert. He gave the cashier $95. He received $3 in change. What did each ticket cost?

1.
2.
3.
4.
5.
6.
7.

Decimals: Comparing

Name ____________________

Compare. Use $>$, $<$, or $=$ for each ◯

39.00162 ◯ 39.00126

39.00162
39.00126

$6 > 2$ so $39.00162 > 39.00126$

To compare 2 decimals:

1. Line up the decimal points.
2. Compare digits from *left to right* in their corresponding place value.

---

1. 3,527.796205 ◯ 3,257.796205
2. 0.065517 ◯ 0.065551
3. 563.09999 ◯ 563.099999
4. 0.00013 ◯ 0.01013
5. 6,223.67212 ◯ 6,223.76212
6. 0.0044561 ◯ 0.044561
7. 39.034763 ◯ 39.34763
8. 222.0202202 ◯ 222.0202202
9. 4,135.331535 ◯ 4,135.335315
10. 3.531031 ◯ 3.513031
11. 623.145889 ◯ 623.14589
12. 7.68799 ◯ 7.68979
13. 15.035893 ◯ 15.035983
14. 481.28764 ◯ 481.27824
15. 881.036487 ◯ 818.036437
16. 0.0112112 ◯ 0.0121121

THINK ABOUT IT!

17. Find a number that is between the two given numbers, 5.29 and 5.3

Decimals: Ordering

Name ______________________

Order from least to greatest.

0.4698, 0.4689, 0.4869

0.4698
0.4689
0.4869
Answer: 0.4689, 0.4698, 0.4869

To compare 2 or more decimals:
1. Line up the decimal points.
2. Starting from the left, compare the digits in their corresponding place value.

TIP: *Comparing two at a time to find the least or greatest may help.*

---

1. 3.45572, 3.45725, 3.45427

2. 224.95003, 224.90553, 224.59993

3. 0.055753, 0.055755, 0.07505

4. 22.21221, 22.21212, 22.22111

5. 4.499441, 4.49944, 4.49994

6. 0.73342, 0.73422, 0.73323

7. 3,671.03, 3,617.03, 3,671.031

8. 456, 0.456, 400.56

9. 314.09901, 314.09091, 314.00991

10. 0.08813, 0.08183, 0.08883

11. 79.06542, 79.60054, 79.60006

12. 5.00387, 5.03008, 5.037708

13. 0.001202, 0.001210, 0.0012002

14. 0.443434, 0.443443, 0.443344

Decimals: Estimating Addition & Subtraction Name ____________

Estimate each sum or difference. Use an appropriate strategy. When rounding, round to the highest place value.

Estimate by rounding.

$$\begin{array}{rcr} 17.9 & \longrightarrow & 20 \\ \underline{-\ 10.3} & \longrightarrow & \underline{-\ 10} \\ \text{Estimate} & & 10 \end{array}$$

Estimate by clustering.

41.2 + 38.8 + 40.3 =
40 + 40 + 40 = 40 x 3 = 120
(The numbers all cluster around 40.)

---

1. 3.27 + 6.75
2. 19.5 + 56.1
3. 248.3 − 105.9
4. 77.86 − 24.35
5. 8.71 + 6.43
6. 68.4 − 21.2
7. 121.5 + 487.8
8. 76.22 − 47.34
9. 9.3 + 10.0 + 10.3 + 11.1
10. 50.4 + 51.1 + 48.9
11. 324.5 − 102.3
12. 100.5 + 97.8 + 101.6 + 99.3
13. 5.98 + 9.82
14. 71.1 + 69.3 + 70.2
15. 985.4 − 648.3
16. 24.8 − 14.9
17. 2.4 + 8.3
18. 16.19 + 2.18
19. 654.9 + 829.4
20. 29.53 − 18.12
21. 8.32 + 8.71 + 7.23
22. 72.4 − 39.8
23. 211.9 + 385.4
24. 54.2 − 28.9

THINK ABOUT IT!

25. Your grocery bill is \$47.82. If you give the clerk a \$50 bill, **about** how much money do you expect in change?

Decimals: Adding & Subtracting

Name____________________________

Find each sum or difference. ***Watch the sign!***

33.4 – 3.82

```
  33.40  ←
–  3.82
  29.58
```

1. Line up the decimal points
2. Add zeros if necessary.
3. Add or subtract as with whole numbers.

TIP: *Remember to bring your decimal point down into your answer.*

---

1. 6,183.62 − 5,812.89
2. 0.883521 + 9.227489
3. 717 − 706.552
4. 5.780125 − 4.612334
5. 103.57782 + 87.40662
6. 25.7 − 0.8761
7. 0.75012 − 0.28334
8. 117 − 105.0023
9. 71.0548 + 18.443
10. 3.956 + 2.41 =
11. .0589 + .278 =
12. 1.14 − .44 =
13. .257 + .768 =
14. 6.48 + 2.14 =
15. .821 − .07 =
16. .456 + 2.8 =
17. 6.788 − 0.2 =
18. .722 + .9 =
19. 3.24 − .51 =
20. 4.44 + 15 =
21. 75.5 − .67 =

Name ____________________

Decimals: Estimating Multiplication & Division

Estimate each product or quotient. Use an appropriate strategy. When rounding, round to the highest place value.

Estimate by rounding.

8.2×9.1

$\downarrow \quad \downarrow$

$8 \times 9 = \mathbf{72}$

Estimate by choosing compatible numbers.

$42.5 \div 6.3$

$\downarrow \quad \downarrow$

$42 \div 6 = \mathbf{7}$

---

1. $11.4 \times 12.9 =$
2. $8.73 \times 9.24 =$
3. $49.42 \div 6.8 =$
4. $27.2 \times 9.7 =$
5. $39.61 \div 5.3 =$
6. $7.9 \times 5.1 =$
7. $119.1 \div 29.2 =$
8. $7.5 \times 8.4 =$
9. $73.8 \div 8.1 =$
10. $36.3 \times 9.7 =$
11. $41.7 \div 19.4 =$
12. $32.5 \times 81.4 =$
13. $35.7 \div 11.4 =$
14. $7.2 \times 9.9 =$
15. $119.3 \div 23.1 =$
16. $4.9 \times 1.4 =$
17. $63.2 \div 9.1 =$
18. $17.8 \div 6.2 =$
19. $22.2 \times 9.5 =$
20. $28.9 \div 6.7 =$
21. $28.2 \times 31.3 =$
22. $55.6 \div 8.2 =$
23. $9.45 \times 6.73 =$
24. $19.23 \times 4.6 =$
25. $29.5 \div 4.5 =$
26. $61.5 \div 11.4 =$
27. $14.5 \times 3.3 =$

Decimals: Multiplying & Dividing

Name ______________________

Find each product.

62.8 × .93

62.8	*1 decimal place*	1. Multiply as you would whole numbers.
× .93	+ *2 decimal places*	2. The number of decimal places in the product is the sum of the decimal places in the factors.
1884		
56520		
58.404	*3 decimal places*	

TIP: *Remember, do* ***not*** *line up the decimal points when setting up your problem.*

---

1. 0.6×0.8
2. $0.9 \times .27$
3. 18.3×0.67
4. 7.2×5.4
5. 8.4×0.003
6. 0.28×28.1
7. 0.53×17.1
8. 1.89×0.375

Find each quotient.

3.9 ÷ .13

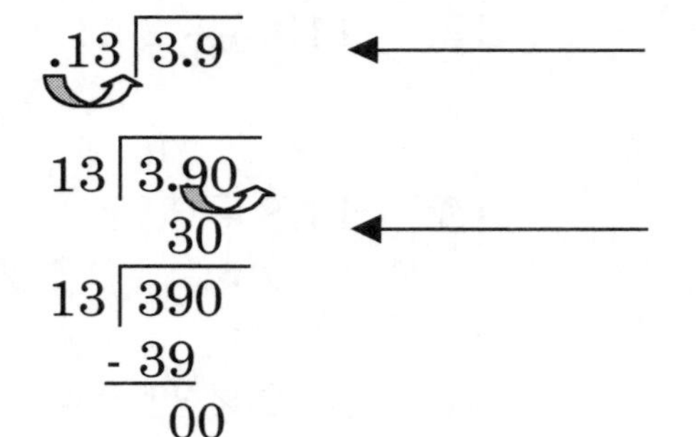

1. Change the divisor to a whole number by moving the decimal point to the right.
2. Move the decimal point in the dividend the same number of spaces. Add zeros if necessary.
3. Divide.

---

9. .4 | 82
10. 3.5 | 2.38
11. 1.4 | 121.8
12. 8 | 0.0092
13. 149.73 ÷ .23
14. 2.004 ÷ .2
15. 165.2 ÷ 8.26
16. 90.36 ÷ 25.1

Decimals: Scientific Notation

Name ____________________

Part I. Write each number in scientific notation.

6,310,000.
$\underline{6.31} \times 10^{\underline{6}}$

1. Move the decimal point to change your number (n) so that $1 \leq n < 10$.

0.000095

$\underline{9.5} \times 10^{\underline{-5}}$

2. Count how many places you moved the decimal point. Place that number as your power of 10 (*positive* if you moved to the left, *negative* if you moved to the right).

---

1. 820
2. 0.0084
3. 2,400
4. 6,324
5. 0.00003
6. 6,784,000
7. 86,000
8. 0.00093
9. 51,000

Part II. Write each number in standard form.

3.4×10^5
340,000

6.8×10^{-4}
.00068

1. Look at the exponent on the 10.
2. If it is *positive*, move the decimal point that many places to the right.
3. If it is *negative*, move it that many places to the left.

---

10. 2.2×10^5
11. 6.104×10^3
12. 1.25×10^{-3}
13. 9×10^8
14. 9.17×10^{-6}
15. 1.96×10^{-4}
16. 6.7×10^{-5}
17. 298.3×10^2
18. 7.4×10

THINK ABOUT IT!

19. Place the following in order from the smallest to the largest.

5.1×10^7 3.2×10^{-3} 9.7×10^6

Decimals: Problem Solving

Name ______________________

Solve each problem.

1. Carolyn spent $19.25 on posters at the music store. Each poster cost $2.75. How many posters did she buy?

2. Ray is a cross country runner. If he runs an average of 36 miles per week, **about** how many miles does he run each day?

3. Sarah earned $5.75 an hour at her after school job. Monday, Wednesday, and Friday she worked from 3 p.m. to 7 p.m. On Tuesday and Thursday she worked from 3 p.m. to 5:30 p.m. How much money did she earn each week?

4. Susan had to replenish some of her supplies. She bought 2 stamp pads at $1.79 each, an eraser for $0.49, and 3 packs of index cards for $0.99. How much did she spend in all? How much change did she receive from a $20 bill?

5. For her art class, Mrs. Lamb bought 15 paint brushes at $3.45 each, an easel for $17.95, and 5 jars of paint at $1.98 each. How much money did she spend altogether?

6. Mrs. Powell worked a total of 45.5 hours last week. She worked 8.75 hours on Monday, 9.2 hours on Tuesday, 10.3 hours on Wednesday, and 8.25 hours on Thursday. How many hours did she work on Friday?

1.
2.
3.
4.
5.
6.

Name ______________________

Algebra: Writing & Evaluating Algebraic Expressions

Part I: Write an algebraic expression for each phrase or problem.

7 more runs scored than the Marlins

Let ***r*** *represent the number of runs scored by the Marlins. The words* ***more than*** *suggest addition,* ***product*** *suggests multiplication, etc.*

r + 7

TIP: *Be sure to identify whether you should add, subtract, multiply or divide.*

---

1. seven more than t
2. the sum of r and 2
3. eight less than p
4. the difference of g and 4
5. c divided by 7
6. three times a
7. the quotient of b and 2
8. the product of a number and 7
9. 6 points less than Ben's score
10. the sum of 9 and c

Part II: Evaluate each expression if a = 10, b = 5, c = 3, and d = 2.

$a \div b$
$10 \div 5 = 2$

1. Substitute the numbers for the variables.
2. Evaluate.

---

11. $ab - 15$
12. $b - c + 2d$
13. $4c + b$
14. abc
15. $c^2 - 2$
16. ab^2
17. $(ab)^2$
18. $3a + 2d - 4c$
19. $\frac{a}{d} - c$

Algebra: Writing Algebraic Equations **Name** ______________________

Translate each problem into an algebraic equation.

A number minus 6 equals 12. Let n = number. $n - 6 = 12$

TIP: *Remember that there is an equals sign in every equation.*

---

1. A number plus 14 equals 25.

2. A number minus 12 is 25.

3. The product of 5 and y added to 3 is 33.

4. 3 more than 5 times the number of cats is 18.

5. The quotient of a number and 6 is 5.

6. The sum of 6 and the product of 7 and a number is 75.

7. The cost of a soda plus 10 cents tax is $2.09.

8. 5 less than the product of 7 and x is 58.

9. The sum of 8 and the quotient of c and 7 is 11.

10. $12 more than 3 times Jason's wages if $450.

11. 14 less than twice the number of dogs is 48.

12. The quotient of c and 5, less 4, equals 10.

13. 53 is 1 more than twice the number of students in a class.

14. 51 is 4 times the number of books, less 13.

THINK ABOUT IT!

15. Write a problem that you might solve using the following equation:

$$6x - 2 = 3$$

Algebra: Problem Solving

Name ______________________

Solve each problem.

1. Translate each phrase into an algebraic expression. Be careful!
   a. five is less than a number
   b. five less than a number
   c. five less a number

2. Shirley's salary is currently **s** dollars a month. Write an expression for the new salary if:
   a. she receives a raise of $100 a month.
   b. her salary is lowered by $50 a month.
   c. her salary is tripled.

3. Write two word phrases for each expression.
   a. 12d
   b. 850 – n
   c. $\frac{x}{5}$
   d. y + 14

4. A library's fine for an overdue book is 25¢. Added to that charge is an additional 5¢ each day the book is overdue. What would be the fine for a book that is returned 6 days late?

5. Suppose a = 268 and b = 3,712.
   a. Evaluate ab – ba.
   b. Will the answer above change if the values of a and b are changed? Why or why not?

1.
2.
3.
4.
5.

Name ____________________

Algebra: Solving Addition & Subtraction Equations

Solve and check each equation.

$x - 15 = 29$
$x - 15 \underline{+ 15} = 29 \underline{+ 15}$
$x = 34$
Check: $34 - 15 = 29$
$29 = 29$ ✔

1. Look at what has been done to the variable.
2. Undo it using the inverse (opposite) operation on both sides of the equation.
3. To check, replace the variable with your solution.

---

1. $d + 32 = 70$
2. $x - 36 = 12$
3. $t - 31 = 30$
4. $2.5 + r = 4$
5. $g - 3.5 = 1.25$
6. $n + 240 = 300$
7. $5.6 + x = 6.3$
8. $n - 305 = 225$
9. $708 = n + 300$
10. $7.09 = 3 + y$
11. $12.3 = 14 - b$
12. $s - 3.2 = 6$
13. $x + 5.25 = 7.55$
14. $64 + x = 146$
15. $7.9 + x = 12.3$
16. $n + 89 = 134$
17. $x - 89 = 176$
18. $m - 76 = 158$
19. $x + 77 = 394$
20. $21 = x - 43$
21. $89 = c + 44$
22. $14 + r = 26$
23. $x + 8.9 = 9.7$
24. $r - 2.3 = 7.9$

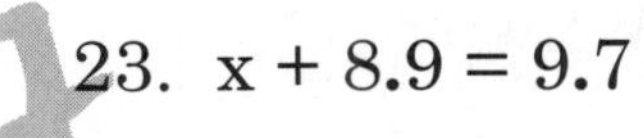

Name ______________________

Algebra: Solving Multiplication & Division Equations

Solve and check each equation.

$27n = 213$
$27n \underline{\div 27} = 213 \underline{\div 27}$
$n = 19$
Check: $27 \times 19 = 213$
$213 = 213$✔

1. Look at what has been done to the variable.
2. Undo it using the inverse (opposite) operation on both sides of the equation.
3. To check, replace the variable with your solution.

---

1. $8n = 128$
2. $64 = 16x$
3. $12n = 96$
4. $n \div 81 = 1.4$
5. $5x = 225$
6. $n \div 36 = 7.5$
7. $\frac{n}{14} = 0.1$
8. $\frac{n}{72} = 2.4$
9. $\frac{n}{10} = 13.6$
10. $9y = 126$
11. $168 = 4x$
12. $3y = 474$
13. $n \div 4 = 1.7$
14. $384 = 12c$
15. $42 = 2.1y$
16. $\frac{n}{5} = 1.5$
17. $\frac{h}{3} = 1.5$
18. $\frac{x}{2} = 18$
19. $n \div 12 = 12$
20. $0.4x = 56$
21. $18 = 0.9s$
22. $r \div 5 = 17$
23. $6.1c = 42.7$
24. $n \div 38 = 67$

Algebra: Solving More Equations Name ______________________

Solve and check the following equations.

1. $x + 15 = 46$
2. $6y = 17.94$
3. $n - 11 = 54$
4. $r \div 1.5 = 24$
5. $x - 13 = 12$
6. $5.25c = 21$
7. $17 = r + 9$
8. $18n = 3.6$
9. $22 = t - 3.3$
10. $\frac{x}{1.8} = 7.2$
11. $\frac{n}{5} = 16.3$
12. $36 = d + 17$
13. $2.6x = 9.62$
14. $30 + y = 48$
15. $3.4m = 8.5$
16. $n - 55 = 146$
17. $1.2n = 2.76$
18. $m - 17 = 26$
19. $14 + x = 49$
20. $8.34y = 25.02$
21. $16 + r = 37$
22. $\frac{c}{2.7} = 31$
23. $\frac{m}{17} = 38$
24. $h \div .3 = 18$
25. $746 = n - 3.7$
26. $63n = 945$
27. $127 + n = 250$
28. $1.9h = 17.1$
29. $2.17 + n = 4.19$
30. $2.1n = 147$

Algebra: Solving Two-Step Equations Name ____________

Solve and check the following equations.

$3n - 2 = 13$

$3n - 2 \underline{+ 2} = 13 \underline{+ 2}$

$3n = 15$

$3n \underline{\div 3} = 15 \underline{\div 3}$

$n = 5$

When solving a two-step equation, undo each operation.

1. First, undo the addition and subtraction.
2. Then, undo the multiplication and division.

$\underline{3 \cdot 5} - 2 = 13$

$15 - 2 = 13$

$13 = 13$ ✔

To check, replace the variable with the solution.

---

1. $6n - 3 = 21$
2. $\frac{x}{5} + 2 = 6$
3. $2y + 7 = 15$
4. $7x + 6.4 = 6.4$
5. $6n - 12 = 78$
6. $\frac{b}{7} - 13 = 12$
7. $5x - 50 = 35$
8. $\frac{x}{10} - 2.5 = 7.5$
9. $9r - 7 = 74$
10. $\frac{h}{7} + 1 = 4$
11. $\frac{n}{2} + 11 = 16$
12. $3n - 5 = 4$
13. $4n + 8 = 32$
14. $4r - 9 = 7$
15. $2e + 5 = 6$
16. $\frac{f}{8} + 3 = 27$
17. $5y + 9 = 24$
18. $2x + 3 = 9$
19. $7y + 12 = 40$
20. $6x - 14 = 22$
21. $62 = 7m + 13$

Algebra: Problem Solving

Name ____________________

Write an equation to help you solve each problem.

1. A number is decreased by 11, and the result is multiplied by 2. The final answer is 20. What is the number?	1.
2. The quotient of x and 9 increased by 3 is 6.	2.
3. Two more than twice the number of books is 8. How many books are there?	3.
4. The sum of x and 5 decreased by 1 is 6.	4.
5. The product of 5 and a number minus 6 is 2 less than the difference of 25 and 4. What is the number?	5.
6. The quotient of the number of bananas and 4 is 3. How many bananas are there?	6.
7. The product of a number and 15 is 180. What is the number?	7.
8. The quotient of a number and 5 increased by 6 is 11. What is the number?	8.
9. The number of receivers on a football team is three times the number of quarterbacks. If there are nine receivers on a team, how many quarterbacks are there?	9.

Integers: Meaning of Integers

Name ______________________

Write an integer for each exercise.

An ***integer*** is the set of whole numbers and their opposites.

If –15 represents a withdrawal of $15, write an integer for a withdrawal of $42.
Answer: –42

TIP: *A plus sign is not needed to write a positive integer; however, a negative sign is required for a negative.*

---

1. A loss of 12 pounds
2. 6 inches taller
3. 34 degrees below zero
4. A loss of 11 yards
5. A profit of $50
6. A gain of 2 pounds
7. The opposite of 15
8. The opposite of –20
9. 4 strokes over par
10. 3 inches shorter
11. An altitude of 2,000 ft
12. 100 ft below sea level
13. 7 units to the left of zero on a number line
14. 16 units to the right of zero on a number line
15. 3 strokes under par
16. A gain of 6 points
17. A loss of $210
18. An increase of 12 degrees
19. A deposit of $100
20. A gain of 25 yards

THINK ABOUT IT!

21. Graph the set of numbers on the number line { -5, -3, 0, 2 }.

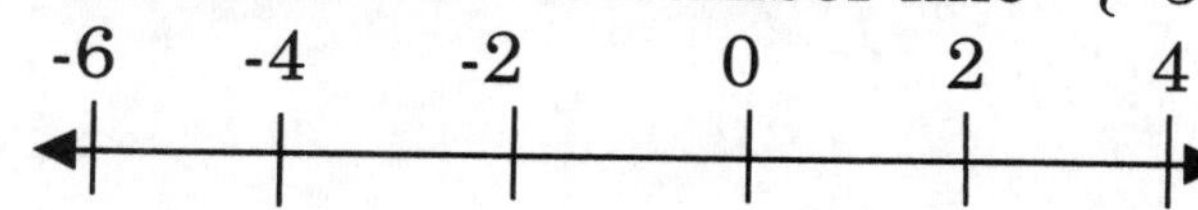

Integers: Absolute Value

Name ____________________

Find each absolute value.

The ***absolute value*** of a number is its distance from zero. The following symbol is used when asked to find the absolute value: | |. (Two straight lines surrounding the number.)

| –10 | = 10 –10 is 10 places from zero so its ***absolute value*** is 10.
| 28 – 11 | = 17 Subtract the numbers first, then find the ***absolute value.***

---

1. | 4 |	2. | –5 |	3. | 11 |	4. | –3 |
5. | 0 |	6. | 6 |	7. | –8 |	8. | 12 |
9. | 23 |	10. | –9 |	11. | –45 |	12. | 33 |
13. | –28 |	14. | –51 |	15. | 61 |	16. | –73 |
17. | 23 – 5 |	18. | 12 + 7 |	19. | –14 |	20. | 38 |
21. |12|–|–8|	22. | 0 + 6 |	23. | –15 |	24. | 65 + 59 |

THINK ABOUT IT!

25. Give the opposite of each problem.

A. | –10 | B. – | 5 | C. | 32 |

Integers: Comparing and Ordering

Name ____________________

Part 1: Compare. Use <, >, or = for each ○.

When comparing integers, the integer that is further to the right on the number line has the *greater value.*

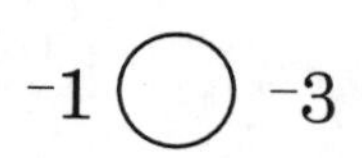

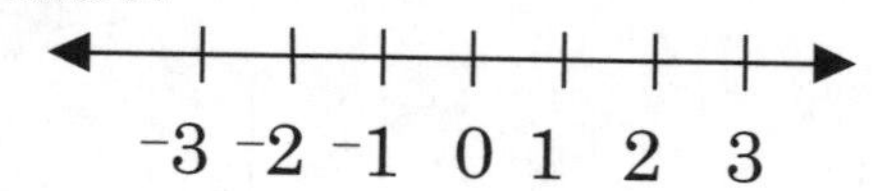

-1 is further to the right, so -1 > -3.

---

1. -5 ○ 0
2. -8 ○ -5
3. 2 ○ -2
4. 10 ○ -70
5. 40 ○ -40
6. -35 ○ -35
7. -10 ○ -24
8. 68 ○ -50
9. 6 ○ -16
10. 20 ○ -30
11. -65 ○ -45
12. -32 ○ -12

Part II: Order from greatest to least.

13. -1, -5, 1, 0, -8, -4
14. 0, -7, 3, 7, -2
15. 14, -5, 1, -1, 0, -9
16. -20, -60, 0, 110, 60, -140
17. 20, -20, 10, -10, 50, -30
18. 60, -30, -10, 0, 40, -40
19. -75, 85, 90, -25, 50
20. -18, -20, -5, -10, -17

Integers: Multiplying & Dividing

Name ____________________

Find each product or quotient.

When the signs are the ***same*** (both positive or negative) the answer will be ***positive.***	When the signs are ***different*** (one positive and one negative) the answer will be ***negative***.
$^-7 \bullet {}^-6 = 42$ $140 \div 2 = 70$	$^-120 \div 4 = {}^-30$ $6 \bullet {}^-15 = {}^-90$

1. $^-3 \bullet {}^-8$
2. $^-35 \div {}^-7$
3. $^-5 \bullet {}^-5$
4. $^-65 \div 5$
5. $^-8 \bullet 2$
6. $^-35 \bullet {}^-3$
7. $100 \div {}^-2$
8. $7 \bullet {}^-3$
9. $^-15 \bullet 3$
10. $^-63 \div 9$
11. $^-28 \div {}^-4$
12. $0 \bullet {}^-21$
13. $240 \div {}^-4$
14. $6 \bullet {}^-3$
15. $^-5 \bullet {}^-30$
16. $144 \div {}^-12$
17. $8 \bullet 15$
18. $^-36 \div {}^-12$
19. $(4 \bullet {}^-8) \bullet {}^-2$
20. $({}^-49 \div 7) \bullet 8$
21. $^-6 \bullet (5 \bullet {}^-2)$
22. $18 \div ({}^-6 \bullet {}^-3)$
23. $({}^-8 \bullet 5) \bullet 10$
24. $({}^-12 \bullet {}^-3) \bullet {}^-2$

Integers: Adding Integers

Name ______________________

With the Same Sign

The sum of two positive integers is positive; the sum of two negative integers is negative.

$^-5 + {^-12} = {^-17}$

$18 + 25 = 43$

With Different Signs

Subtract their absolute values and use the sign of the greater absolute value.

$^-10 + 23 = 13$

$|\ 23\ | - |\ {^-10}\ | =$

$\downarrow \qquad \downarrow$

$23 \quad - \quad 10 = 13$

Find each sum.

1. $^-4 + 8$
2. $14 + 16$
3. $(^-6) + (^-6)$
4. $(^-9) + (^-5)$
5. $^-19 + 11$
6. $^-26 + 40$
7. $52 + (^-18)$
8. $^-42 + 39$
9. $28 + (^-42)$
10. $^-43 + (^-12)$
11. $^-15 + (^-11)$
12. $47 + 12$
13. $^-86 + 18$
14. $75 + (^-5)$
15. $^-49 + (^-32)$
16. $^-200 + 92$
17. $^-130 + (^-2)$
18. $23 + (^-23)$
19. $32 + (^-18 + {^-15})$
20. $^-12 + 16 + (^-6)$
21. $^-15 + 15 + 20$

THINK ABOUT IT!

22. Complete each statement. Use ***sometimes, always,*** or ***never.***

a. The sum of two negative integers is ____________ positive.

b. The sum of a negative number and a positive one is ________ negative.

c. The sum of a positive number and its opposite is ____________ zero.

Integers: Subtracting Integers

Name ____________________

Find each difference.

To subtract integers, add the opposite.

$10 \underline{- (-5)}$	$-15 \ \underline{-\ 4}$	$-3 \underline{- (-9)}$
$10 \ \underline{+5} = 15$	$-15 + \underline{-4} = -19$	$-3 \ \underline{+9} = 6$

TIP: *Never change the sign of the first integer, just the second.*

---

1. $4 - 7$
2. $-5 - 3$
3. $-8 - 2$
4. $10 - 17$
5. $8 - (-5)$
6. $13 - 9$
7. $-3 - 24$
8. $53 - 68$
9. $0 - (-16)$
10. $-18 - 0$
11. $-4 - (-7)$
12. $4 - (-7)$
13. $68 - (-10)$
14. $62 - (-29)$
15. $-61 - 15$
16. $74 - (-61)$
17. $-41 - 37$
18. $-51 - 21$
19. $-6 - (-6)$
20. $-89 - 1$
21. $36 - (-14)$

THINK ABOUT IT!

22. Evaluate $x - y$ if $x = -6$ and $y = 10$. Then evaluate $y - x$ for the same values of the variables. Use your results to make a conjecture about the relationship between $x - y$ and $y - x$. Does your conjecture hold true if $x = 6$ and $y = 9$?

Integers: Solving Equations

Name ____________________

Solve and check each equation. Remember to use the inverse operation to solve the variable.

$$w + {}^{-}8 = 12$$
$$w + {}^{-}8 \underline{- (^{-}8)} = 12 \underline{- (^{-}8)} \qquad \textit{Subtract } {}^{-}8 \textit{ from both sides.}$$
$$w = 20$$

---

1. $^{-}4 + x = 12$
2. $m - 10 = {}^{-}8$
3. $^{-}7 = y + 12$
4. $3x = {}^{-}15$
5. $^{-}8x = {}^{-}368$
6. $^{-}2a = 8$
7. $98 = \frac{a}{-4}$
8. $\frac{x}{12} = {}^{-}6$
9. $\frac{c}{-4} = 10$
10. $56 = \frac{y}{-3}$
11. $^{-}240 = \frac{h}{2}$
12. $^{-}345 = t + 56$
13. $6 + x = 98$
14. $x - 45 = {}^{-}2$
15. $3m = {}^{-}168$
16. $x + 9 = {}^{-}34$
17. $75 + p = {}^{-}100$
18. $0 = 6r$
19. $x + 25 = 15$
20. $^{-}25c = 125$
21. $^{-}570 = 3t$

THINK ABOUT IT!

22. Write an equation for the problem below; then, solve it.

If you decrease a number d by $^{-}4$, the result is 10. Find d.

Integers: Problem Solving

Name ____________________

Solve each problem.

1. Points **w**, **x**, **y**, and **z** are different points on a number line. Use the following clues to put the four integers in order of least to greatest:
   - **z** is the least of the integers.
   - Point **y** is a positive integer.
   - Points **w** and **z** are the same distance from 0.
   - Point **y** is closer to **z** than it is to **x**.

2. Rita opened a checking account with a balance of $150. She wrote a check for $77.
   a. Write an addition sentence to represent this situation.
   b. How much money remained in the account?

3. During a space shuttle launch, a maneuver is scheduled to begin at T minus 85 seconds which is 85 seconds before liftoff! The maneuver lasts 2 minutes. At what time will this maneuver be complete?

4. Find each product.
   a. $(^-3)(4)(5)$
   b. $(^-3)(^-4)(5)$
   c. What is the sign of the product $(^-3)(^-4)(^-5)$?
   d. Write a rule for determining the sign of the product of **three** nonzero integers.

5. The water level in a tank decreased 10 cm in 5 minutes. If the tank drains at a steady rate, what is the change in the water level each minute?

1.

2.

3.

4.

5.

Statistics: Reading Charts & Tables

Name ____________________

Use the following table and information below to answer the questions.

The senior class wants to sell a commemorative t-shirt or sweatshirt. To determine which to sell, and what color, the class officers survey the 450 students. The results are shown in the table below.

Preference	Number of Votes
t-shirt	320
sweatshirt	130
red	56
white	149
blue	97
black	150

1. List the colors in order from most popular to least.

2. Based upon the color results, is there a clear preference for color choice?

3. If you were a class officer, what would you recommend about the color choice? Explain your answer.

4. What percent of the students prefer a t-shirt? a sweatshirt?

5. Based upon your answer to # 4 above, should the officers choose the t-shirt or sweatshirt for their sale?

Statistics: Graphing Skills

Name ____________________

After studying the graph below, answer the questions that follow.

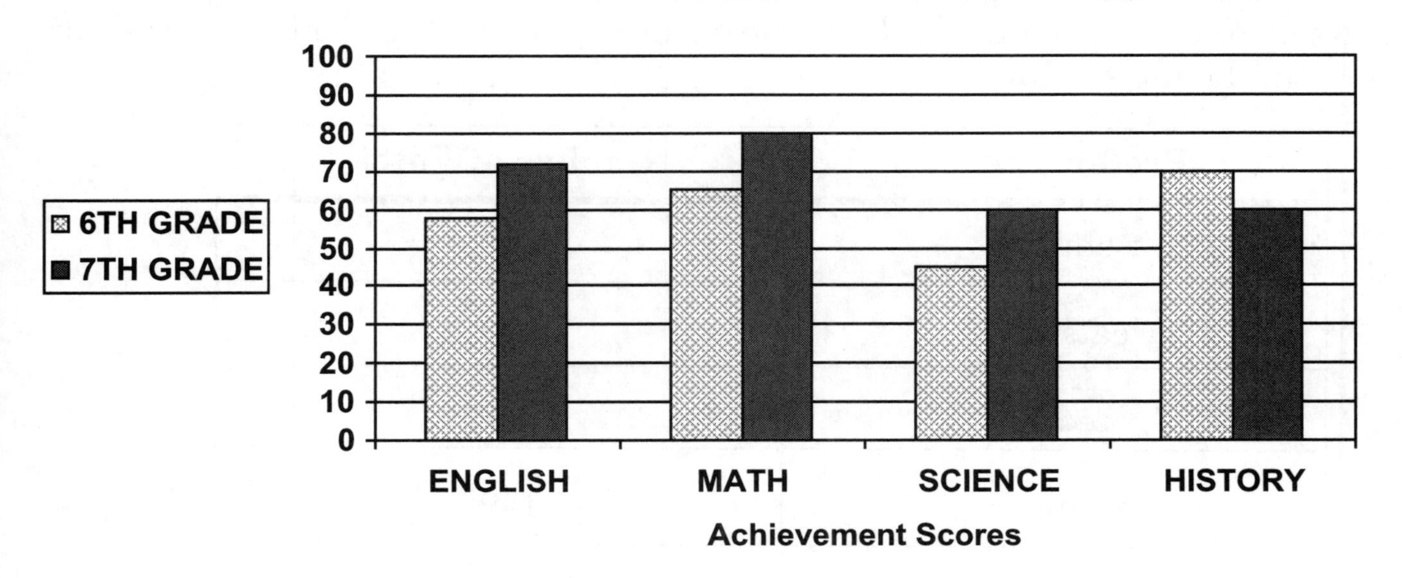

1. In which subject did the seventh graders score the highest?
   O A. English O B. Math O C. Science O D. History

2. Which subject had the lowest overall score?
   O A. English O B. Math O C. Science O D. History

3. What was the total of the achievement scores in all subjects for the sixth graders?
   O A. 272 O B. 253 O C. 510 O D. 238

4. What was the difference in scores between the sixth and seventh graders in history?
   O A. 15 O B. 10 O C. 5 O D. 20

5. How much higher did the seventh graders score in all subjects on the achievement test than the sixth graders?
   O A. 34 O B. 44 O C. 29 O D. 39

6. What is the closest estimate of the achievement scores by the sixth graders in English and Math?
   O A. 130 O B. 100 O C. 60 O D. 250

Statistics: Predicting Outcomes

Name ______________________

Use the samples given to predict the outcomes.

A sample can be used to make a *prediction* about a larger number of items. Cross multiply to find the value of "x."

4 red marbles out of 20
30 red marbles out of x.

$$\frac{4}{20} = \frac{30}{x} \qquad x = 150$$

---

1. Sample: 6 out of 120 are green.
   Actual: 50 out of x are green. ______________________

2. Sample: 8 out of 32 are blue.
   Actual: 40 out of x are blue. ______________________

3. Sample: 4 out of 36 are yellow.
   Actual: 50 out of x are yellow. ______________________

4. Sample: 10 out of 100 are orange.
   Actual: 250 out of x are orange. ______________________

5. Sample: 2 out of 30 are purple.
   Actual: 60 out of x are purple. ______________________

6. Sample: 20 out of 25 are pink.
   Actual: 80 out of x are pink. ______________________

7. A bag has 3 colors of gumballs in it. There are 250 gumballs in all. A sample of 20 gumballs contains 10 blue, 4 red, and 6 yellow. How many gumballs of each color do you predict are in the bag?

   ____

8. In a classroom box of several crayons, 1 out of 3 crayons is blue. There are other colors in the box as well. In a sample of 45 crayons, how many of the blue crayons would you expect to find? ______________________

Name ______________________

Statistics: Mean, Median and Mode

Find the mean, median, and mode of each set of data. Round to the nearest tenth.

5, 8, 8, 7, 8, 5

Mean (or average): Add the numbers and divide by the total number in set.
$(5 + 8 + 8 + 7 + 8 + 5) \div 6 = 41 \div 6 = 6.8$ ***Mean*** = 6.8

Median (middle number): Place the numbers in order. Find the middle number. If there is not one middle number, average the two in the middle.
5, 5, 7, 8, 8, 8 The two middle numbers are 7 and 8. $(7 + 8) \div 2 = 7.5$
Median = 7.5

Mode (most frequent): Find the number that occurs most frequently.
Mode = 8

1. 1, 5, 8, 5, 3, 6, 4, 3, 2, 7

2. 3, 4, 7, 5, 6, 7, 2, 8

3. 90, 92, 94, 90, 91, 95, 94, 98

4. 13, 20, 17, 18, 12, 17, 15

5. 60, 45, 70, 50, 60, 42, 42, 60, 45

6. 56, 75, 65, 57, 76, 64, 65, 66

7. 8.9, 8.0, 9.0, 9.4, 9.2, 9.3

8. 144, 150, 133, 144, 125, 100

9. 1,755; 1,780; 1,755; 1,805; 1,805

10. 300, 100, 250, 600, 150, 400, 250

Statistics: Problem Solving

Name ____________________

Solve each problem.

1. The low temperatures for the first week of winter were 20, 28, 31, 32, 29, 26, 20.
   a. Find the mean, median and mode.
   b. Which number would you use to emphasize how cold it was?

2. The manager of Sporting Shoes keeps a record of the sizes of each athletic shoe sold. Which number is probably most useful: the mean, median or mode? Please explain.

3. Create a set of data with five numbers for each of the following situations:
   a. The mean is greater than the median.
   b. The mean, median, and mode are the same.
   c. The mean is not equal to one of the numbers in the set.

4. A reporter for the school newspaper randomly stopped 50 students leaving gym class and asked them if they thought there was enough time between classes. Forty-two students replied that there was not enough time. Do you think that the survey reflected the opinions of all students? Why or why not?

5. A movie advertisement reads "Thousands of people rate this movie as the BEST MOVIE OF THE YEAR." What kind of information would you want to know before believing that this movie really is the best one of the year?

1.
2.
3.
4.
5.

Geometry: Classifying Angles

Name ____________________

Classify each angle as <u>acute</u>, <u>right</u>, <u>obtuse</u>, or <u>straight</u>.

Acute Angle $< 90^\circ$

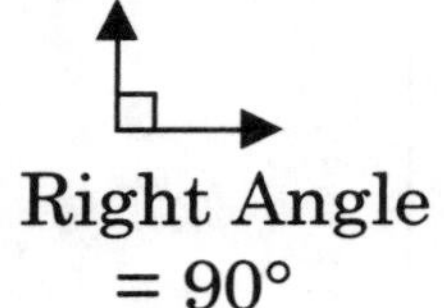

Right Angle $= 90^\circ$

Obtuse Angle $90^\circ < x < 180^\circ$

Straight Angle $= 180^\circ$

---

1. 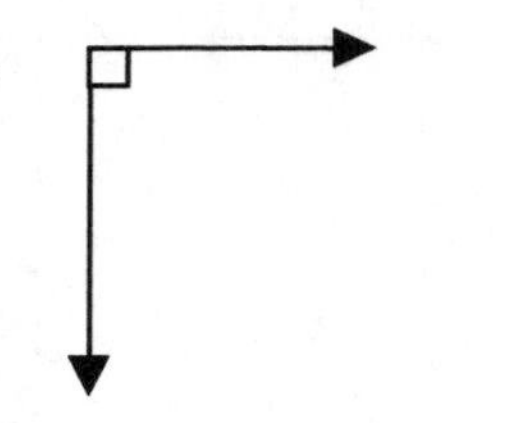

2. 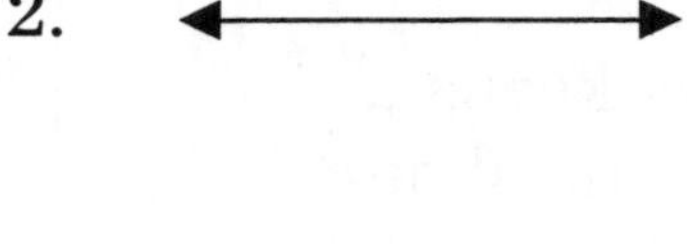

3. 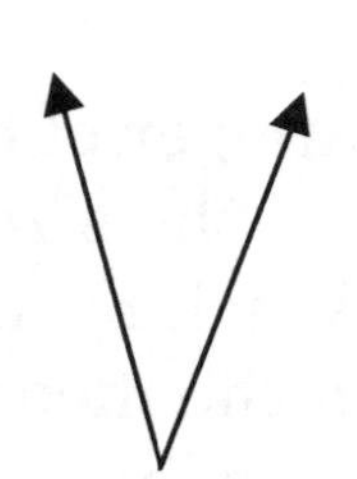

4. 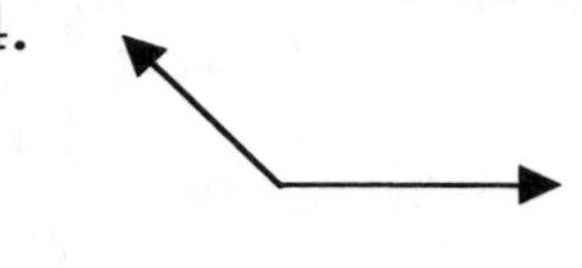

5. 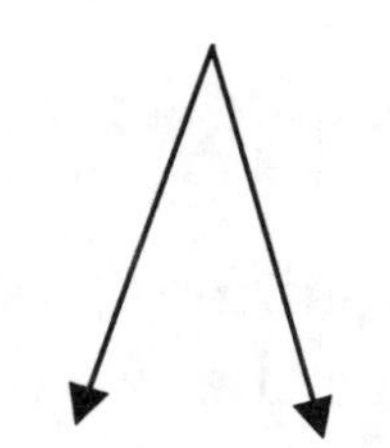

6. 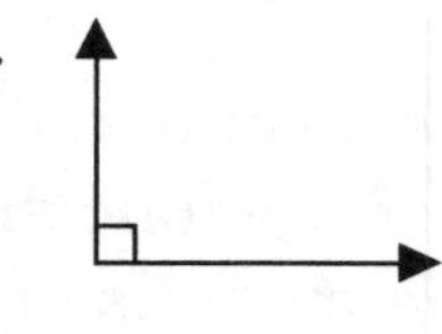

7. 42° angle

8. 91° angle

9. 180° angle

10. 112° angle

11. 17° angle

12. 90° angle

THINK ABOUT IT!

13. Without measuring, match each angle to the appropriate measurements.

A. 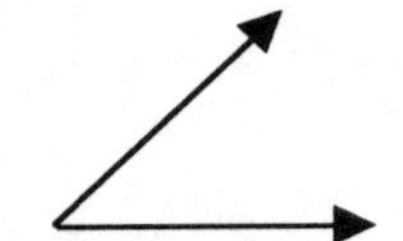

B. 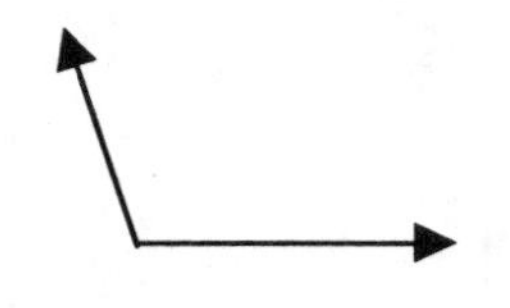

C. 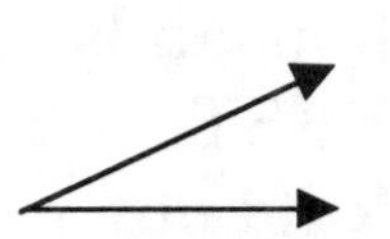

D. 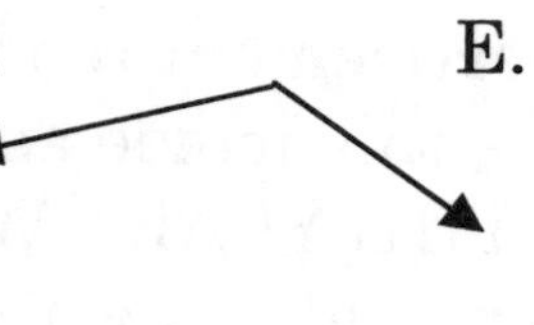

E. 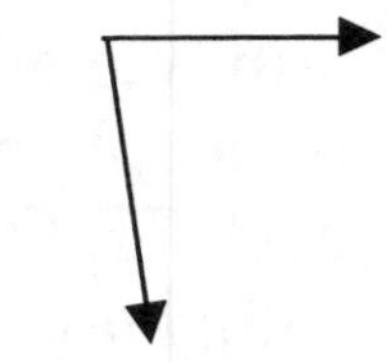

a. 40° b. 50° c. 85° d. 100° e. 110°

Geometry: Classifying Triangles

Name ____________________

Classify each triangle by its sides and angles.

By sides:

Equilateral

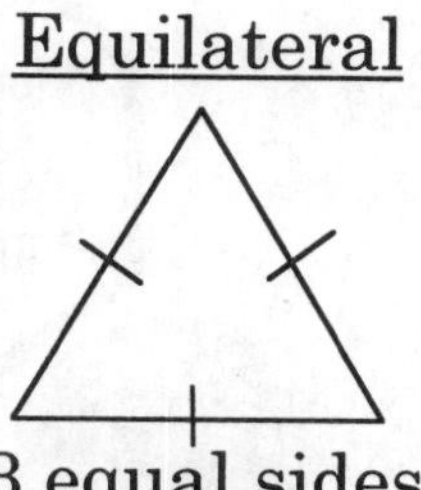

3 equal sides

Isosceles

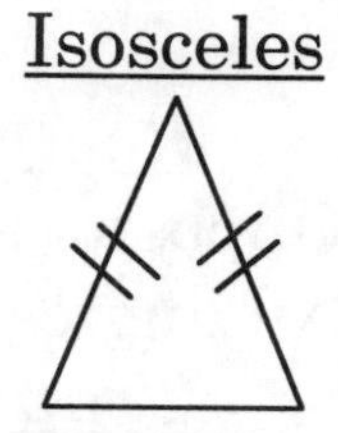

2 equal sides

Scalene

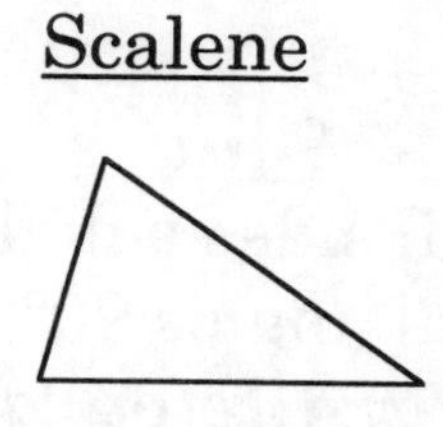

no equal sides

By angles:

Acute

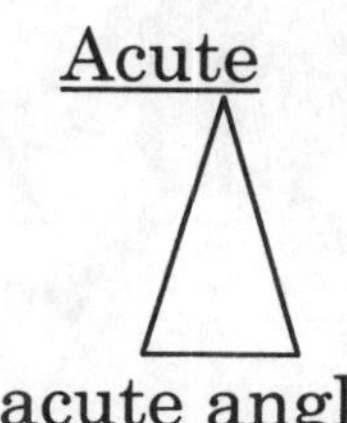

3 acute angles

Obtuse

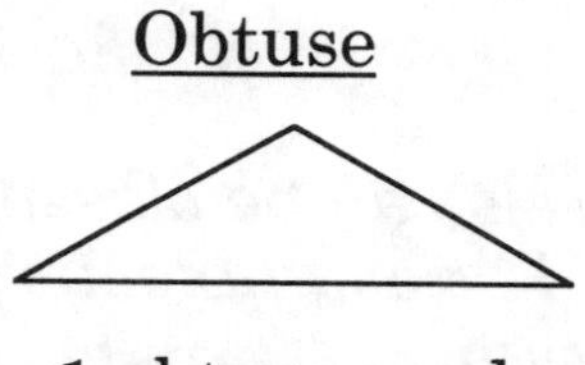

1 obtuse angle

Right

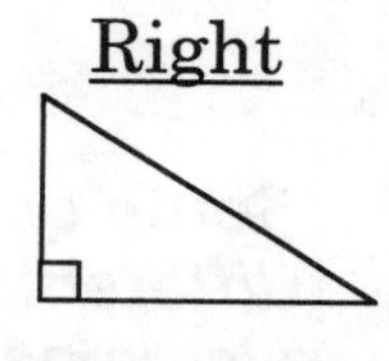

1 right angle

1.

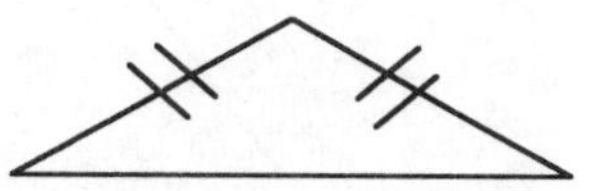

2.

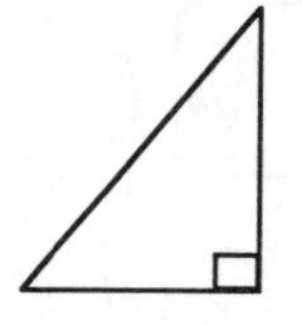

3.

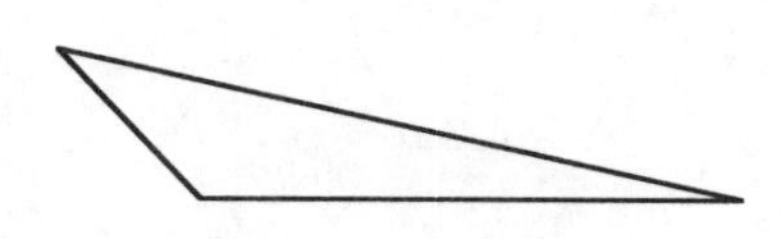

4.

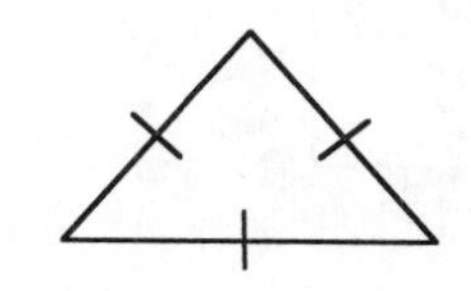

5.

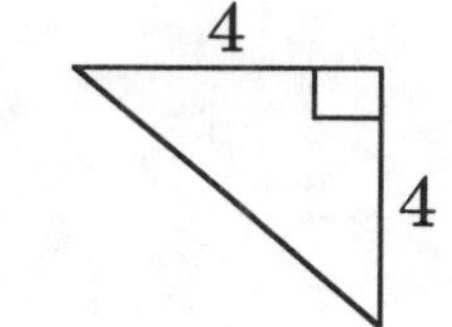

6.

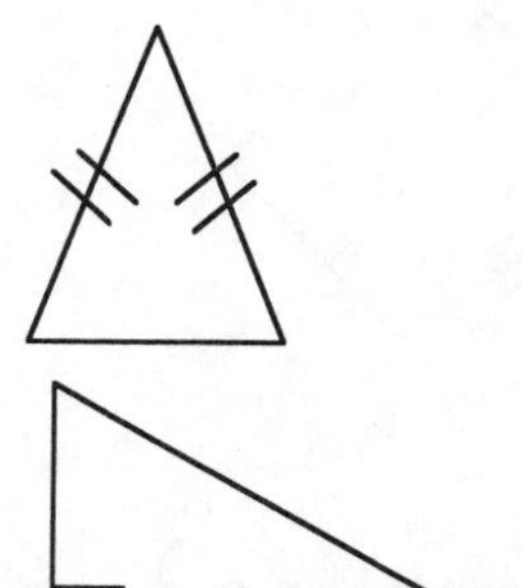

7.

8.

THINK ABOUT IT!

9. Draw a triangle that fits the description of each problem below.
   a. a scalene obtuse triangle
   b. an equilateral acute triangle
   c. an isosceles right triangle

Geometry: Classifying Quadrilaterals   Name ____________

Name every quadrilateral that describes each figure. Then underline the name that best describes the figure.

MATH FACTS

More than 2,000 years ago, the Greek civilization perfected Trigonometry, the study of the angles between, and lengths of an object's sides.

Square
All sides equal
All angles 90°

Parallelogram
Opposite sides parallel

Rectangle
Opposite sides equal
All angles 90°

Trapezoid
1 pair of parallel sides

Rhombus
All sides equal

TIP: *Some quadrilaterals can be classified by two different names. For example, a rectangle is also a parallelogram.*

---

1. 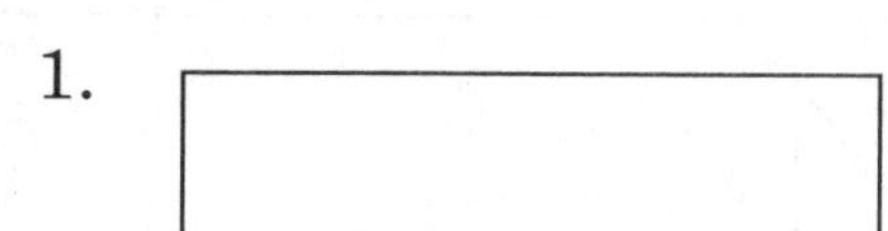

2. 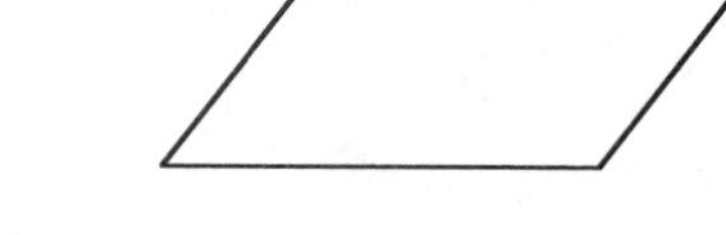

3. 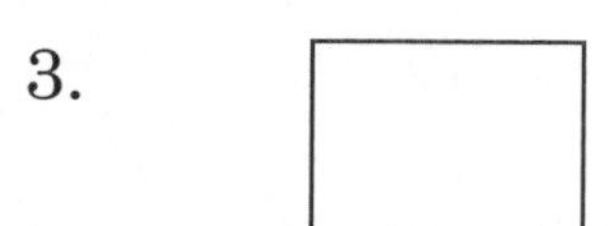

4. 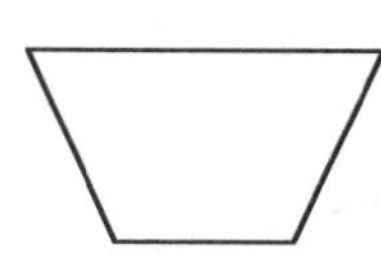

5. 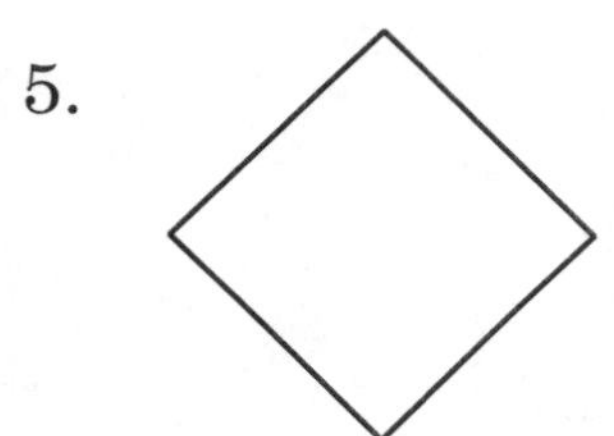

6. 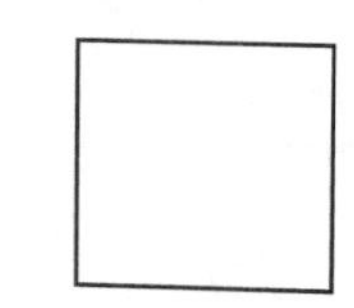

THINK ABOUT IT!

7. Answer the questions below.
   a. Which quadrilaterals can have four right angles?
   b. Is every square a rectangle? Explain.
   c. Are all parallelograms also squares?
   d. Write a sentence that tells how a trapezoid is different from a parallelogram.

Geometry: Angle Sums

Name ______________________

Find the missing measure in each triangle or quadrilateral.

The sum of the measures of the angles of any triangle is 180°.

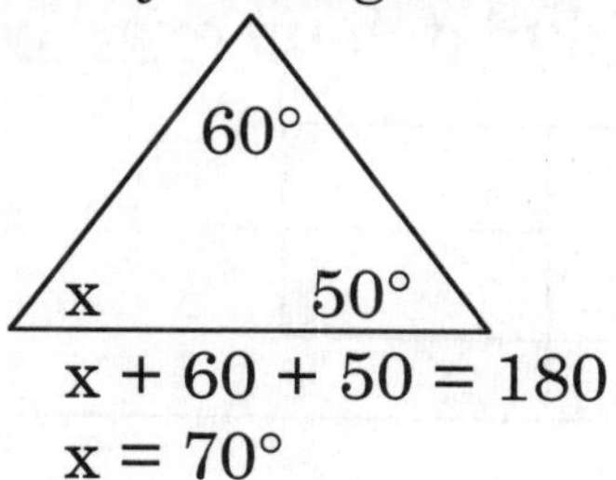

$x + 60 + 50 = 180$

$x = 70°$

The sum of the measures of the of the angles of any quadrilateral is 360°.

x 95°

110° 88°

$x + 95 + 88 + 110 = 360$

$x = 67°$

1.

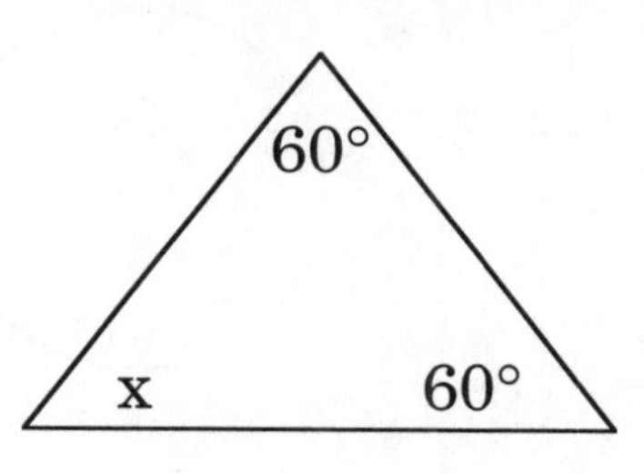

2.

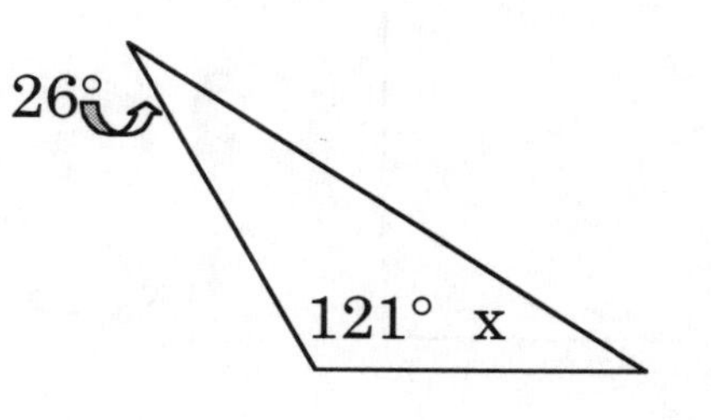

3.

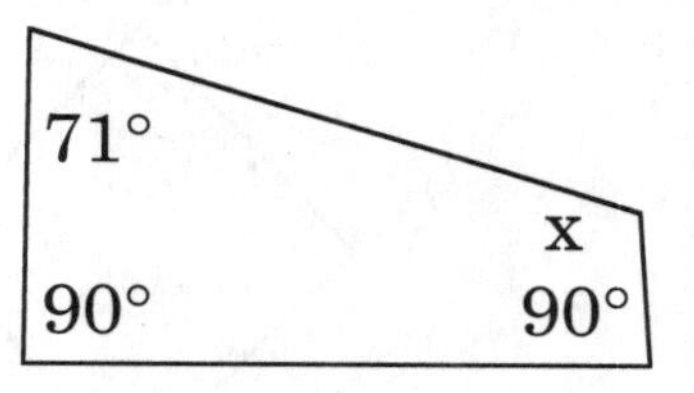

4.

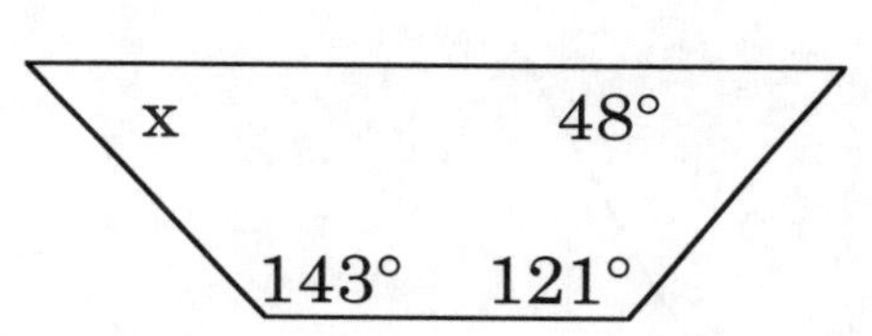

5.

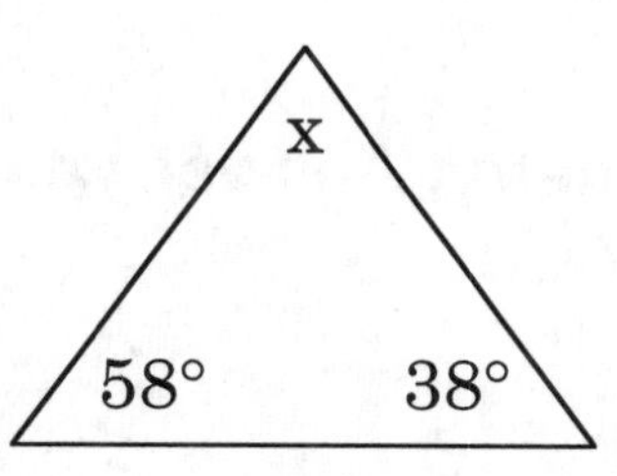

6.

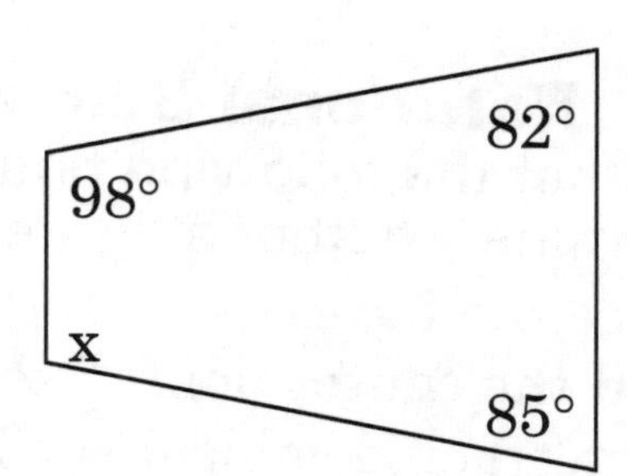

THINK ABOUT IT!

7. One angle of a triangle has a measure of 48°. The other two angles have equal measures. Find the measure of the other two angles.

Geometry: Symmetry

Name ______________________

Part I: Line Symmetry

Draw all lines of symmetry in each figure. If the figure has none, write none.

A line of symmetry cuts a figure in half so that if you were to fold the figure along that line, the sides would match up exactly.

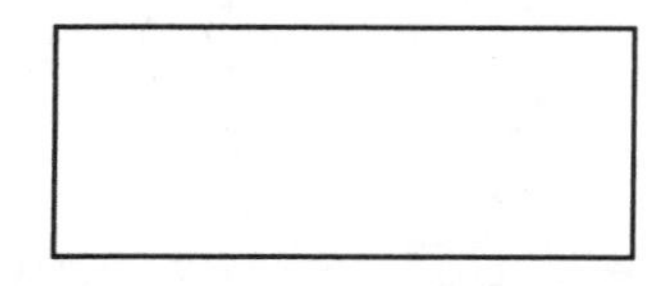

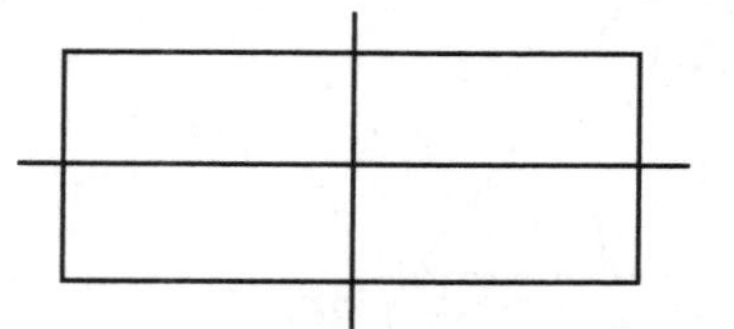

1.

2.

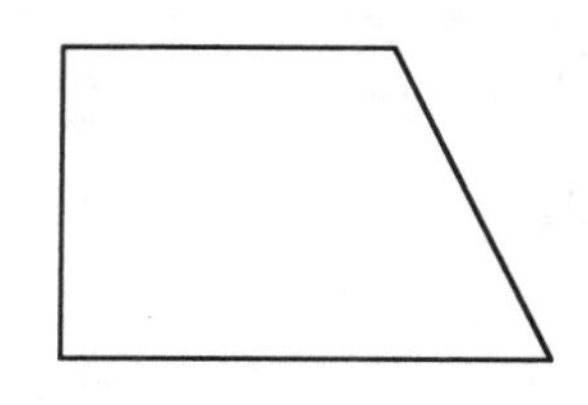

3.

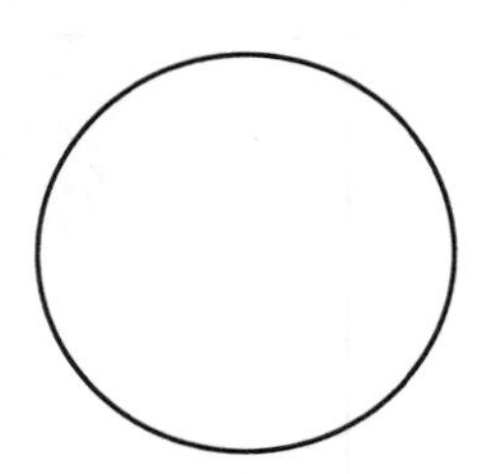

4.

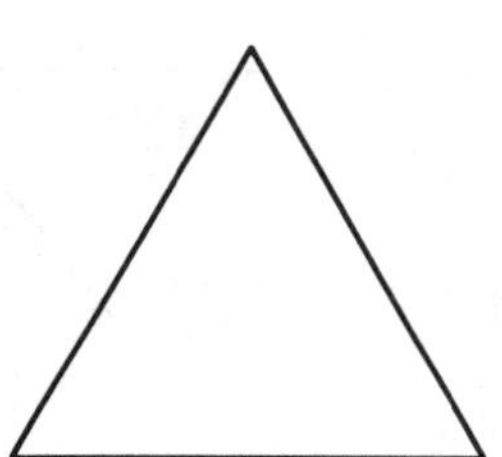

5.

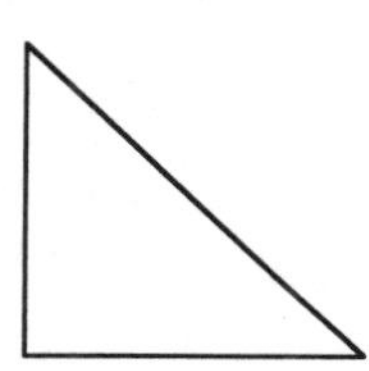

6.

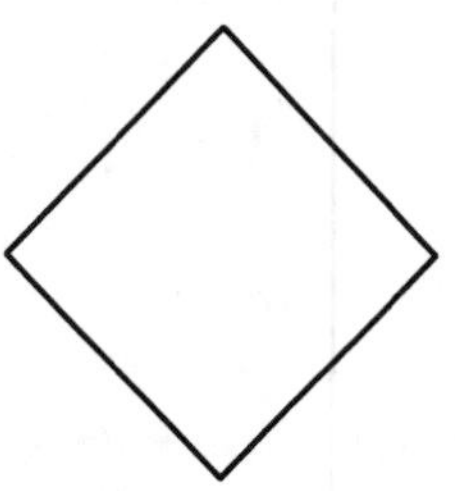

Part II: Rotational Symmetry

Tell whether the following figures have rotational symmetry. Answer ***yes*** or ***no.***

To determine whether a figure has rotational symmetry:

1. Trace the figure.
2. Hold the center point fixed and turn the tracing.
3. If the tracing coincides with the original figure before it can be turned a full turn (360°), then it has rotational symmetry.

7.

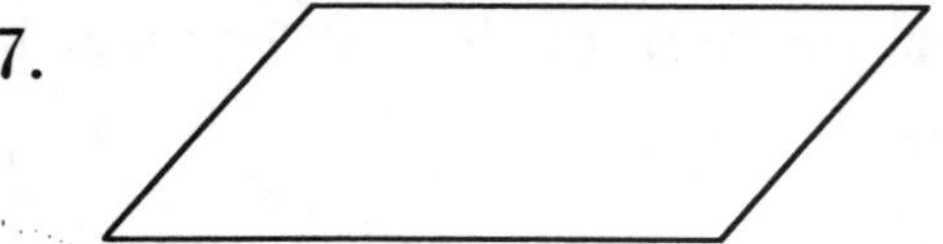

8.

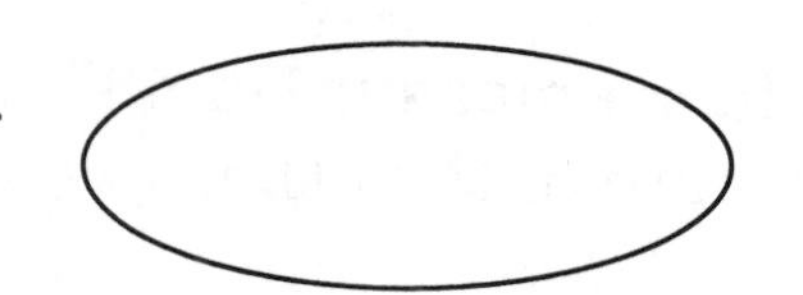

9.

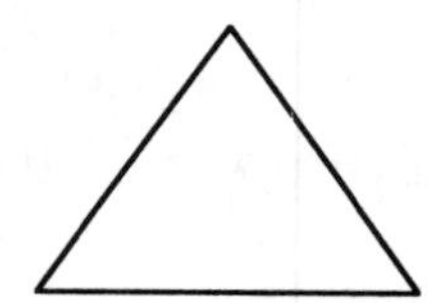

Geometry: Coordinate Planes

Name ____________________

Part I: Naming Points on a Graph. Name the coordinates of the points. Use the coordinate system below. Remember: the first coordinate is the distance from 0 on the x-axis and the second coordinate is the distance from 0 on the y-axis.

1. A
2. B
3. C
4. D
5. E
6. F

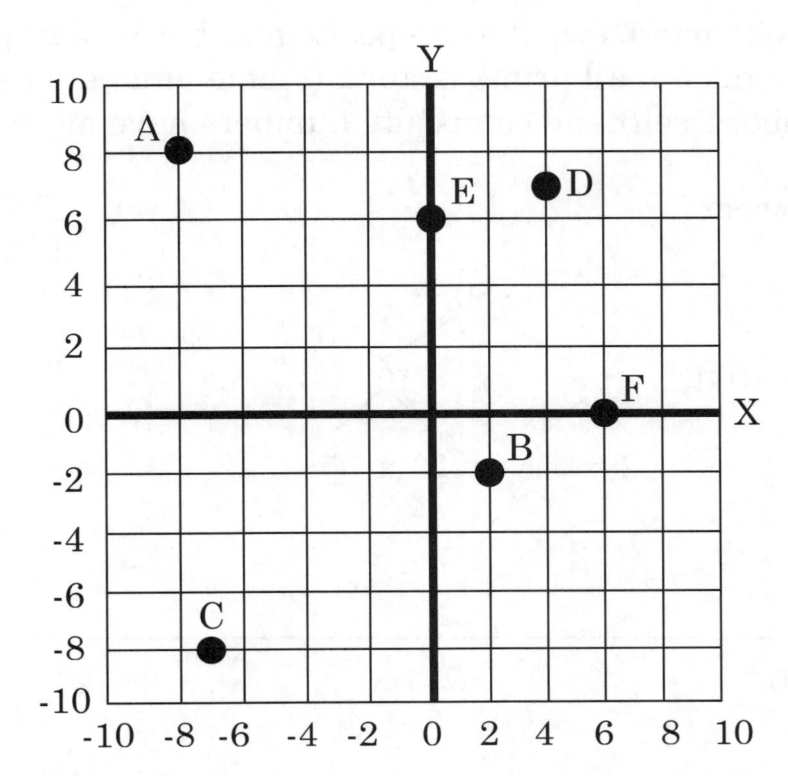

Part II: Graphing Points. On graph paper, draw a coordinate plane. Then graph each set of points. Label each point.

7. G (-4, 4)
8. H (2, 4)
9. I (2, -2)
10. J (0, 0)
11. K (-6, -3)
12. L (-5, 0)
13. M (0, 4)
14. N (-2, 7)

THINK ABOUT IT!

15. List five ordered pairs which lie on the same horizontal line. What generalization can you make about the y-coordinates of points that lie on the same horizontal line? Repeat this exercise for a vertical line.

Fractions: Prime Factorization

Name ______________________

Use a factor tree to find the ***prime factorization*** of each number. Use exponents when possible.

Prime factorization of a composite number is taking that number and expressing it as a product of all prime factors. (Prime numbers have only two factors {1 and the number itself} and composite numbers have more than two factors.)

Factor tree:

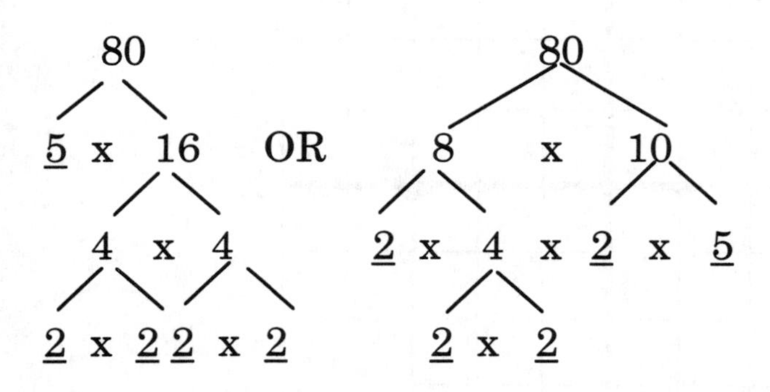

Prime Factorization

$2 \cdot 2 \cdot 2 \cdot 2 \cdot 5 = 2^4 \cdot 5$

---

1. 90

2. 121

3. 175

4. 236

5. 72

6. 270

7. 300

8. 51

Fractions: Greatest Common Factor Name ____________________

Find the ***Greatest Common Factor*** (GCF) of each pair of numbers.

45, 54

45: 1, 3, 5, ⑨, 15, 45
54: 1, 2, 3, 6, ⑨, 18, 27, 54
GCF = 9

1. List all factors of each number.
2. Find the greatest factor common to both numbers.

TIP: *Finding the Greatest Common Factor is especially helpful in the reduction of fractions*

---

1. 45, 75
2. 100, 30
3. 39, 91
4. 12, 78
5. 8, 9
6. 16, 28
7. 18, 54
8. 18, 45
9. 120, 72
10. 132, 108
11. 18, 27
12. 26, 52
13. 6, 8, 12
14. 10, 15, 20
15. 18, 42, 60

Fractions: Lowest Common Multiple

Name ____________________

Find the ***Lowest Common Multiple*** (LCM) for each pair of numbers.

6, 9

6: 6, 12, (18), 24
9: 9, (18)
LCM = 36

1. List some multiples for each number.
2. Find the smallest multiple common to both numbers.

TIP: *The Lowest Common Multiple is very helpful when adding and/or subtracting fractions.*

---

1. 3, 15
2. 60, 12
3. 60, 80
4. 300, 18
5. 625, 30
6. 16, 176
7. 6, 16
8. 8, 9
9. 24, 36
10. 12, 18, 28
11. 32, 80, 96
12. 56, 64
13. 11, 22, 33
14. 5, 18, 45
15. 13, 14

Name ____________

Fractions: Fractions as Decimals, Decimals as Fractions

Part I: Express each fraction as a decimal. Round to the nearest hundredth.

$\frac{3}{4}$

$3 \div 4 = .75$

To change a fraction to a decimal, divide the numerator by the denominator.

---

1. $\frac{6}{25}$
2. $\frac{9}{20}$
3. $\frac{8}{250}$
4. $\frac{13}{50}$
5. $\frac{7}{40}$
6. $\frac{20}{30}$
7. $\frac{14}{16}$
8. $\frac{18}{13}$
9. $\frac{8}{11}$
10. $\frac{12}{200}$
11. $\frac{56}{50}$
12. $\frac{7}{12}$

Part II: Express each decimal as a fraction in lowest terms.

$.45 = \frac{45}{100} = \frac{9}{20}$ $.03 = \frac{3}{100}$

---

13. .5
14. .64
15. .09
16. .83
17. .48
18. 2.5
19. .540
20. .075
21. .18
22. 6.04
23. .675
24. 3.05

Fractions: Comparing and Ordering

Name ______________________

Part 1: Write <, >, or = for each ◯.

$\frac{3}{4}$ ◯ $\frac{5}{6}$ To compare two fractions find their cross products and compare.

$^{18}\frac{3}{4}$ ◯ $\frac{5}{6}^{20}$ 18<20, so $\frac{3}{4} < \frac{5}{6}$

---

1. $\frac{7}{10}$ ◯ $\frac{4}{5}$
2. $\frac{2}{10}$ ◯ $\frac{1}{6}$
3. $\frac{7}{9}$ ◯ $\frac{3}{5}$
4. $\frac{14}{25}$ ◯ $\frac{3}{4}$
5. $\frac{8}{24}$ ◯ $\frac{20}{60}$
6. $\frac{3}{9}$ ◯ $\frac{2}{8}$
7. $\frac{5}{12}$ ◯ $\frac{4}{9}$
8. $\frac{1}{3}$ ◯ $\frac{3}{5}$
9. $\frac{18}{24}$ ◯ $\frac{10}{18}$
10. $\frac{4}{6}$ ◯ $\frac{5}{9}$
11. $\frac{1}{3}$ ◯ $\frac{2}{4}$
12. $\frac{2}{3}$ ◯ $\frac{8}{12}$
13. $\frac{6}{15}$ ◯ $\frac{4}{10}$
14. $\frac{7}{12}$ ◯ $\frac{12}{42}$
15. $\frac{5}{14}$ ◯ $\frac{2}{6}$

Part II: Order from least to greatest.

TIP: *Try changing the fractions to decimals, then compare them.*

16. $\frac{2}{9}, \frac{3}{5}, \frac{3}{4}, \frac{2}{11}$
17. $\frac{2}{3}, \frac{2}{7}, \frac{2}{5}, \frac{2}{9}$
18. $\frac{2}{5}, \frac{1}{4}, \frac{1}{2}, \frac{3}{7}$
19. $\frac{5}{8}, \frac{8}{9}, \frac{13}{18}, \frac{2}{3}$
20. $\frac{7}{8}, \frac{4}{5}, \frac{6}{7}, \frac{5}{6}$
21. $\frac{1}{3}, \frac{1}{4}, \frac{1}{5}$

Fractions: Simplifying Fractions

Name ____________________

Express each fraction in simplest form.

$\frac{6}{24}$ To write a fraction in simplest form, divide its numerator and denominator by their GCF and write the resulting fraction. $\frac{6 \div 6}{24 \div 6} = \frac{1}{4}$

TIP: *Remember: A fraction is in lowest terms when the GCF of its numerator and denominator is **1**.*

---

1. $\frac{25}{45}$	2. $\frac{15}{35}$	3. $\frac{81}{90}$	4. $\frac{64}{80}$
5. $\frac{8}{9}$	6. $\frac{20}{30}$	7. $\frac{14}{28}$	8. $\frac{5}{15}$
9. $\frac{9}{51}$	10. $\frac{9}{24}$	11. $\frac{75}{90}$	12. $\frac{54}{56}$
13. $\frac{24}{40}$	14. $\frac{14}{35}$	15. $\frac{4}{30}$	16. $\frac{71}{82}$
17. $\frac{15}{60}$	18. $\frac{85}{255}$	19. $\frac{84}{128}$	20. $\frac{24}{32}$
21. $\frac{333}{900}$	22. $\frac{180}{270}$	23. $\frac{640}{960}$	24. $\frac{312}{390}$

THINK ABOUT IT!

25. Find all the ways the numbers **2, 4, 8,** and **16** can be written in the squares to make a true statement. $\frac{\square}{\square} = \frac{\square}{\square}$

Name ____________________

Fractions: Improper Fractions & Mixed Numbers

Part I: Change each improper fraction to a mixed number in simplest form or a whole number.

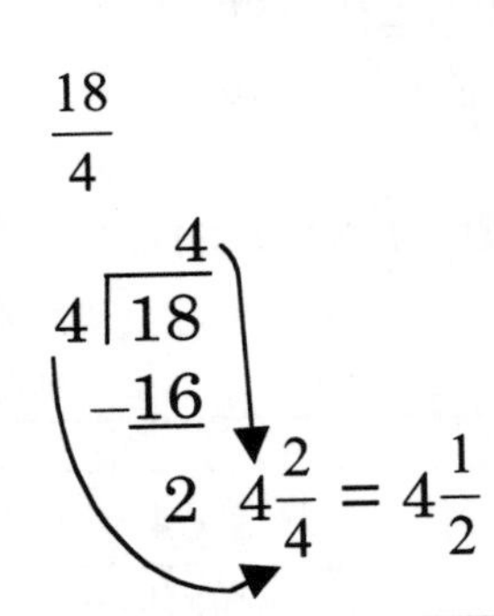

1. Divide the numerator by the denominator.
2. If there is a remainder, put it in fraction form over the divisor.
3. Reduce fraction to lowest terms.

---

1. $\frac{9}{7}$ 2. $\frac{7}{4}$ 3. $\frac{12}{5}$ 4. $\frac{8}{8}$ 5. $\frac{20}{8}$

6. $\frac{17}{5}$ 7. $\frac{6}{2}$ 8. $\frac{21}{9}$ 9. $\frac{26}{6}$ 10. $\frac{23}{7}$

11. $\frac{4}{3}$ 12. $\frac{21}{4}$ 13. $\frac{22}{5}$ 14. $\frac{19}{3}$ 15. $\frac{6}{5}$

Part II: Change each mixed number or whole number to an improper fraction.

$5\frac{1}{3}$

$5 \times 3 + 1 = \frac{16}{3}$

1. Multiply the whole number by the denominator.
2. Add the numerator.
3. Place that number over the denominator.

---

16. $3\frac{4}{5}$ 17. $2\frac{2}{3}$ 18. $2\frac{3}{8}$ 19. $5\frac{3}{5}$ 20. $3\frac{7}{8}$

21. $6\frac{3}{8}$ 22. $7\frac{2}{3}$ 23. $9\frac{2}{5}$ 24. $6\frac{3}{5}$ 25. $8\frac{4}{7}$

26. $2\frac{7}{10}$ 27. $3\frac{5}{12}$ 28. $6\frac{2}{7}$ 29. $3\frac{4}{9}$ 30. $3\frac{6}{7}$

Fractions: Problem Solving

Name ______________________

Solve each problem.

1. Tell why $2^4 \times 15$ is not the prime factorization of 240.

2. Richards Middle School has 525 boys and 600 girls. Express the number of girls as a fraction of the total student population. Write the fraction in simplest form.

3. Karen has won 24 of her 30 golf matches during the golf season.
   a. In simplest form, what fraction of her matches has she won?
   b. In simplest form, what fraction of her matches has she lost?

4. Samuel read 125 pages of his 300-page book.
   a. What fraction of the book has he read?
   b. If Amy has read the same fraction of her 240-page book, how many pages has she read?

5. Could a fraction have more than one simplest form? Explain your answer.

6. The Highland Corner Bakery sold 28 pieces of pecan pie today. If each piece was an eighth of a pie, how many pecan pies did they sell?

7. How many whole notes are there in eight quarter notes?

8. Ellen has $10 worth of quarters and Tim has $10 in half-dollars coins. Who has the greatest number of coins?

1.
2.
3.
4.
5.
6.
7.
8.

Fractions: Adding and Subtracting

Name ______________________

Find each sum or difference. Reduce.

$$\begin{array}{r} \frac{5}{6} = \frac{20}{24} \\ + \frac{3}{8} = \frac{9}{24} \\ \hline \frac{29}{24} = 1\frac{5}{24} \end{array}$$

1. Find the lowest common denominator (LCD).
2. Write the equivalent fractions using the LCD.
3. Add or subtract the numerators. Write the sum or difference over the LCD. Reduce if necessary.

---

1. $\frac{2}{7} + \frac{3}{8}$

2. $\frac{1}{6} + \frac{3}{5}$

3. $\frac{5}{16} - \frac{2}{9}$

4. $\frac{1}{6} + \frac{2}{3}$

5. $\frac{3}{4} - \frac{5}{8}$

6. $\frac{3}{4} - \frac{5}{12}$

7. $\frac{11}{16} + \frac{1}{4}$

8. $\frac{7}{9} - \frac{1}{3}$

9. $\frac{4}{5} - \frac{5}{8}$

10. $\frac{4}{15} + \frac{9}{10}$

11. $\frac{3}{7} + \frac{4}{5}$

12. $\frac{5}{8} - \frac{1}{2}$

13. $\frac{9}{14} + \frac{3}{7}$

14. $\frac{5}{8} - \frac{5}{36}$

15. $\frac{8}{9} + \frac{7}{15}$

16. $\frac{9}{10} - \frac{1}{6}$

Fractions: Adding and Subtracting

Name ____________________

Find each sum or difference. Reduce.

$$\begin{array}{rcrcr} 7\frac{1}{6} & = & 7\frac{2}{12} & = & 6\frac{14}{12} \\ -\,3\frac{1}{4} & = & 3\frac{3}{12} & = & 3\frac{3}{12} \\ \hline & & & & 3\frac{11}{12} \end{array}$$

1. Find the lowest common denominator (LCD).
2. Write the equivalent fractions using the LCD.
3. Rename if necessary.
4. Add or subtract the whole numbers and fractions. Reduce if necessary.

---

1. $3\frac{6}{7} + 4\frac{1}{8}$

2. $4\frac{3}{5} - 2\frac{2}{3}$

3. $8\frac{1}{2} + 2\frac{4}{5}$

4. $5\frac{5}{6} - 4\frac{9}{10}$

5. $2\frac{1}{2} + 4\frac{11}{16}$

6. $7\frac{1}{2} - 1\frac{7}{8}$

7. $8\frac{3}{16} + 2\frac{5}{8}$

8. $8\frac{3}{8} - 7\frac{3}{4}$

9. $4\frac{1}{6} + 8\frac{1}{4}$

10. $6\frac{1}{4} - 3\frac{7}{10}$

11. $3\frac{2}{3} + 2\frac{1}{6}$

12. $4\frac{2}{9} - 1\frac{1}{6}$

13. $\frac{3}{4} + 5\frac{1}{2} + 2\frac{5}{8} =$

14. $12 - 7\frac{7}{10} =$

15. $18\frac{1}{2} + 5\frac{3}{4} + 1\frac{9}{10} =$

16. $25 - (16\frac{1}{5} + 2\frac{4}{5}) =$

17. $7\frac{3}{5} - 2\frac{9}{10} =$

18. $3\frac{3}{10} - 2\frac{2}{5} =$

Name ____________________

Fractions: Multiplying Fractions and Mixed Numbers

Find each product. Reduce.

$1\frac{1}{3} \times 6 =$

$\frac{4}{\cancel{3}_1} \times \frac{\cancel{6}^2}{1} = \frac{8}{1} = 8$

1. Change each mixed number to an improper fraction.
2. Multiply the numerators.
3. Multiply the denominators.
4. Reduce if possible.

TIP: *You can reduce first by dividing a numerator and denominator by a common factor. This is called cross cancellation.*

---

1. $\frac{1}{3} \times \frac{1}{3}$
2. $\frac{2}{9} \times \frac{3}{8}$
3. $\frac{3}{10} \times \frac{2}{3}$
4. $1\frac{3}{4} \times 7$
5. $\frac{3}{5} \times \frac{5}{3}$
6. $4\frac{4}{5} \times 3\frac{3}{4}$
7. $4 \times \frac{5}{12}$
8. $6 \times 2\frac{5}{8}$
9. $\frac{3}{7} \times \frac{5}{6}$
10. $9 \times 3\frac{2}{3}$
11. $\frac{5}{6} \times \frac{3}{5}$
12. $3\frac{1}{2} \times \frac{7}{10}$
13. $\frac{3}{5} \times \frac{10}{21}$
14. $1\frac{4}{7} \times 4\frac{2}{3}$
15. $4\frac{1}{2} \times 1\frac{1}{3}$
16. $3 \times 2\frac{1}{7}$
17. $\frac{4}{5} \times \frac{1}{3} \times \frac{5}{12}$
18. $\frac{5}{8} \times 2 \times \frac{4}{5}$
19. $\frac{6}{7} \times \frac{1}{4} \times \frac{2}{5}$
20. $\frac{5}{8} \times \frac{4}{5} \times \frac{1}{2}$

THINK ABOUT IT!

21. Observe that $5 \times \frac{1}{5} = 1$, $7 \times \frac{1}{7} = 1$, and $10 \times \frac{1}{10} = 1$. What number times $1\frac{1}{2}$ equals 1? times $2\frac{1}{2}$?

Name ______________________

Fractions: Dividing Fractions and Mixed Numbers

Find each quotient. Reduce.

$3\frac{7}{10} \div 2\frac{1}{2}$

$\frac{37}{10} \div \frac{5}{2}$

$\frac{37}{10} \times \frac{2}{5} = \frac{74}{50} = 1\frac{12}{25}$

1. Write the mixed numbers (or whole numbers) as improper fractions.
2. To divide fractions, flip the second one, change it into its reciprocal, and then multiply.
3. Reduce.

TIP: *Remember that a whole number can be written as a fraction by placing it over one.* $4 = \frac{4}{1}$

---

1. $\frac{1}{3} \div \frac{1}{6}$

2. $\frac{3}{4} \div \frac{1}{2}$

3. $\frac{5}{8} \div \frac{1}{16}$

4. $2 \div 1\frac{1}{4}$

5. $5 \div \frac{5}{8}$

6. $\frac{5}{12} \div \frac{3}{16}$

7. $6\frac{1}{4} \div 5$

8. $1\frac{1}{3} \div 2\frac{5}{6}$

9. $2\frac{1}{2} \div 3\frac{3}{4}$

10. $6\frac{2}{3} \div 3\frac{1}{9}$

11. $11\frac{3}{4} \div 5\frac{3}{4}$

12. $\frac{11}{16} \div \frac{3}{16}$

13. $20\frac{1}{8} \div 4\frac{3}{8}$

14. $4\frac{1}{6} \div 3\frac{3}{4}$

15. $\frac{5}{6} \div \frac{5}{8}$

16. $8 \div 1\frac{1}{4}$

17. $7 \div \frac{1}{2}$

18. $4\frac{3}{5} \div 3\frac{1}{5}$

Fractions: Solving Equations

Name ____________________

Solve and check each equation.

$$n - \frac{6}{8} = \frac{2}{3}$$

$$n - \frac{6}{8} + \frac{6}{8} = \frac{2}{3} + \frac{6}{8}$$

$$n = 1\frac{5}{12}$$

1. Look at what has been done to the variable.
2. Undo it by using the inverse (opposite) operation on both sides of the equation.

$$1\frac{5}{12} - \frac{6}{8} = \frac{2}{3}$$

$$\frac{17}{12} - \frac{6}{8} = \frac{2}{3}$$

$$\frac{34}{24} - \frac{18}{24} = \frac{16}{24} = \frac{2}{3} \checkmark$$

3. Check your answer by plugging it back into the equation to see if it makes the equation true.

---

1. $x - \frac{2}{3} = \frac{4}{9}$

2. $x + \frac{3}{4} = \frac{8}{9}$

3. $m - \frac{3}{10} = \frac{5}{8}$

4. $\frac{4}{5}y = 5$

5. $6x = \frac{4}{3}$

6. $c + \frac{3}{4} = \frac{4}{5}$

7. $y - \frac{10}{30} = \frac{2}{5}$

8. $x + \frac{1}{2} = \frac{7}{10}$

9. $1\frac{2}{3}x = \frac{6}{5}$

10. $1\frac{2}{9} = 18h$

11. $\frac{x}{12} = 2\frac{3}{10}$

12. $\frac{3}{7} = x + \frac{2}{5}$

13. $\frac{1}{5} + y = \frac{1}{4}$

14. $\frac{5}{6}x = \frac{7}{12}$

15. $6n = \frac{3}{5}$

Fractions: Problem Solving

Name ____________________

Solve each problem below.

1. The total length of the bicycle race track is $\frac{5}{8}$ mile. The first $\frac{1}{5}$ mile is hilly and the rest is flat. What fraction of the course is flat?

2. Jerry grew a lot during the first half of last year. During the second half, he only grew $\frac{1}{8}$ inch. He grew $3\frac{1}{2}$ inches in all last year. How much did he grow during the first part of the year?

3. The cooking instructions for a turkey recommends roasting the turkey at a low temperature $\frac{3}{4}$ hour for each pound. How long should you cook a $10\frac{1}{2}$ lb. turkey?

4. If a certain steel bar weighs $2\frac{1}{2}$ pounds per foot, what would be the weight of a piece $3\frac{3}{4}$ feet long?

5. An adult ticket to a show cost $2\frac{1}{2}$ times the cost of a child's ticket. If an adult ticket is $3.75, what would be the total cost for 1 adult and 1 child?

6. In one year, 120 students enrolled at a community college. This was $\frac{3}{5}$ of the number of students accepted. How many of those accepted did not enroll?

1.
2.
3.
4.
5.
6.

Measurement: Customary System

Name ____________________

Study the charts below. Then, complete the problems.

Length	*Weight*	*Capacity*
12 in = 1 ft	16 oz = 1 lb	8 fl oz = 1 c
3 ft = 1 yd	2,000 lbs = 1 T	2 c = 1 pt
5,280 ft = 1 mi		2 pt = 1 qt
		4 qt = 1 gal

TIP: *When going from a larger unit to a smaller one, multiply. (Example: miles to feet) When going from a smaller unit to a larger one, divide, (Example: cups to gallons).*

---

1. 70 ft = ______ yd	2. 2 gal = ______ qt	3. 2.5 qt = ______ pt
4. 3 T = ______ lb	5. 80 ft = ______ yd	6. 3 mi = ______ ft
7. 5 c = ______ pt	8. 4 gal = ______ pt	9. 8 yd = ______ ft
10. 7 yd = ______ in	11. 128 oz = ______ lb	12. 3 c = ______ fl oz
13. 5,000 lb = ____ T	14. 16 qt = ______ gal	15. 8 ft = ______ in
16. 4.5 pt = ______ c	17. 42 in = ______ ft	18. 2 qt = ______ c
19. $4\frac{1}{2}$ T = ______ lb	20. 5 lb = ______ oz	21. 1 pt = ______ qt

THINK ABOUT IT!

22. The owner's manual of Harry's car states that it has a capacity of 12 gallons 1 quart of gasoline. Cary's car has a capacity of 12.2 gallons.
    a. Which car has the greater gasoline capacity?
    b. How many more quarts will it hold?

ZIP!

Measurement: The Metric System

Name ______________________

Study the chart below, then complete each sentence.

→ (multiply going right)

× 10 × 10 × 10 × 10 × 10 × 10

kilo-	hecto-	deka-	***meter*** ***liter*** ***gram***	deci-	centi-	milli-

÷10 ÷10 ÷10 ÷ 10 ÷ 10 ÷ 10

← (divide going left)

4.5 m = <u>450</u> cm
$4.5 \times 10^2 = 450$

40 cm = <u>.0004</u> km
$40 \div 10^5 = .0004$

---

1. 5.8 m = _____ cm
2. 567 mg = ______ g
3. 8.1 L = ______mL
4. 8.23 m = _____ cm
5. 13.2 cm = _______ m
6. 6.9 m = ______ mm
7. 329 mL = _______ L
8. 0.62 m = ______ cm
9. 426 cm = _______ m
10. 5.01 g = ______ cg
11. 4,285 mm = _____ m
12. 12.5 m = _______ cm
13. 72.8 km = ______ m
14. 4.3 g = ________ mg
15. 0.32 kg = ______ mg
16. 2,895 L = ______ kL
17. 586 m = _______ km
18. 16.5 cm = _____ km

THINK ABOUT IT!

19. Order the following four lengths from least to greatest.
    0.0021 km   52.35 mm   3.3 cm   0.48 m

Name ____________________

Measurement: Perimeter & Area of Quadrilaterals

Find the perimeter and area of each quadrilateral.

Perimeter**:** distance around a figure

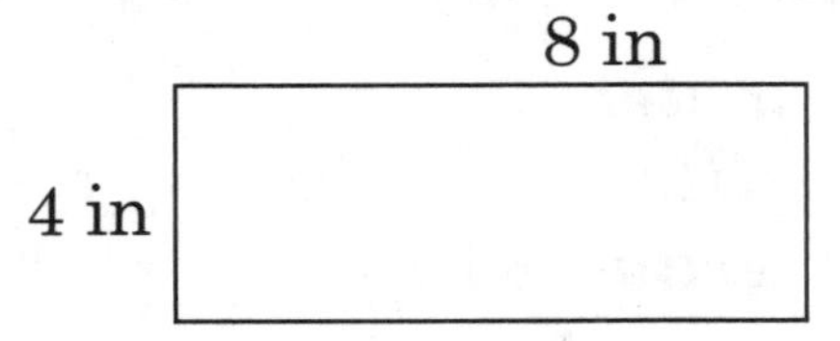

P = 2 × 4 + 2 × 8 = 24 in

Area: measure of what is inside

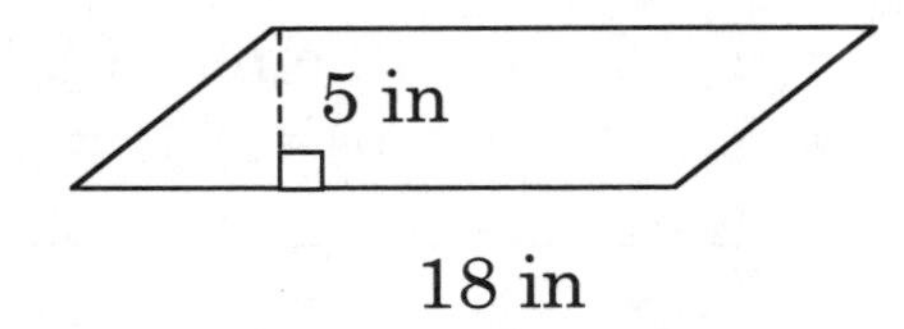

A= b × h
= 18 × 5
= 90 in²

1.

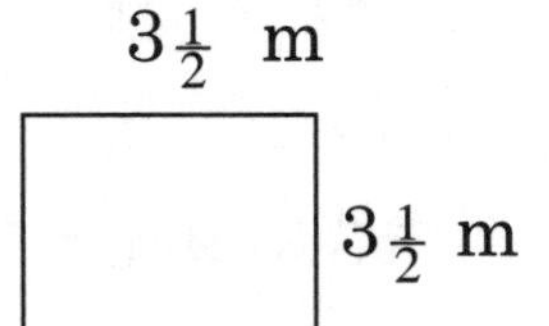

P=
A=

2.

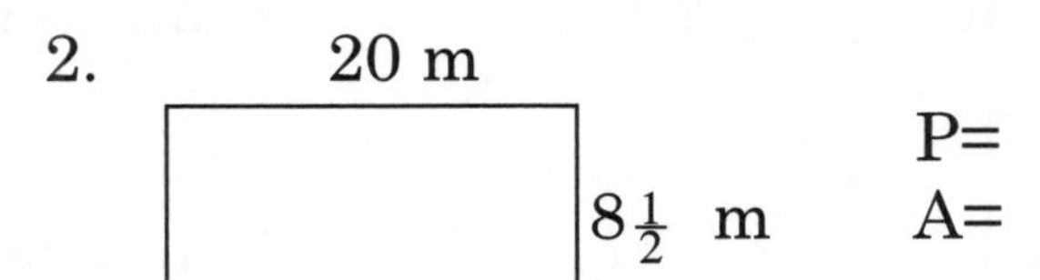

P=
A=

3.

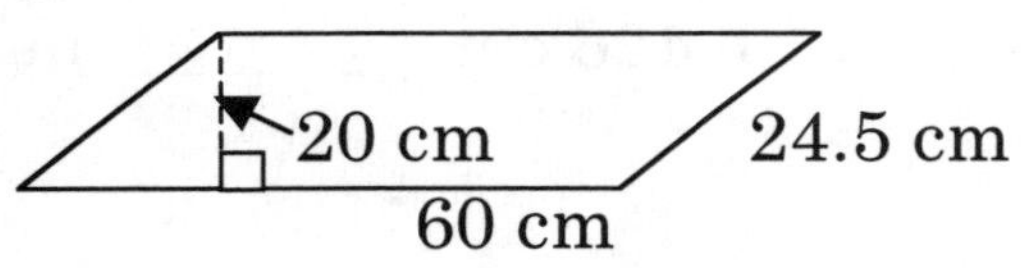

P=
A=

4.

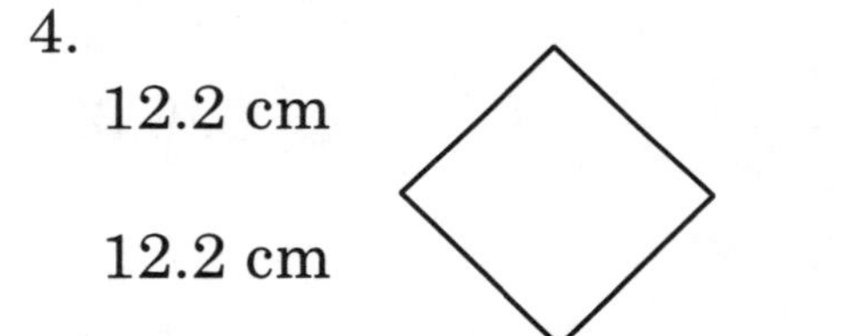

P=
A=

5. 10 m

$8\frac{1}{2}$ m 9 m

P=
A=

6. 30 yd

6.5 yd

P=
A=

7. rectangle:
l = 3.5 ft
w = 1.7 ft

P=
A=

8. square:
side = 2 ft

P=
A=

THINK ABOUT IT!

9. Draw a rectangle equal in area to a 6 x 15 rectangle, but not equal in perimeter to a 6 x 15 rectangle.

Measurement: Squares & Square Roots Name ______________

Part I: Find the square of each number.

When you compute 3×3 or 3^2, you are finding the square of 3.

MATH FACTS

The Greek astronomer, Eudoxus, used trigonomic angle measurements to estimate the size of the earth more than 2,300 years ago!

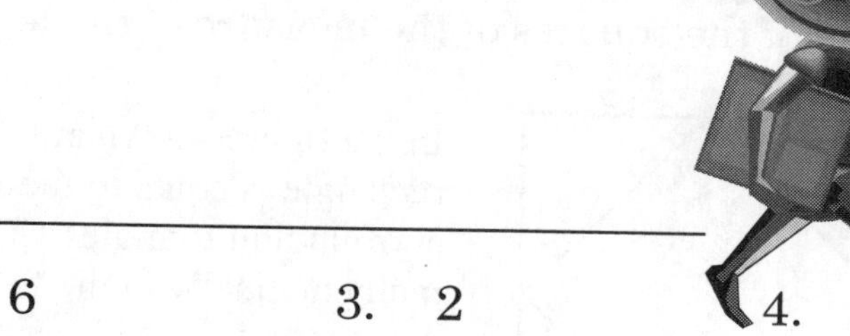

1. 4 2. 6 3. 2 4. 7 5. 1

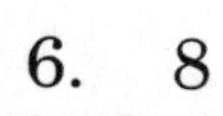

6. 8 7. 12 8. 11 9. 10 10. 9

11. 5 12. 14 13. 15 14. 20 15. 13

Part II: Find each square root.

The symbol that is used to find a non-negative square root is $\sqrt{}$, which is *called a* ***radical sign.***

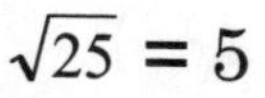

$\sqrt{25} = 5$ Since $5^2 = 25$, 5 is the ***square root*** of 25.

16. $\sqrt{49}$ 17. $\sqrt{16}$ 18. $\sqrt{81}$ 19. $\sqrt{36}$ 20. $\sqrt{4}$

21. $\sqrt{100}$ 22. $\sqrt{64}$ 23. $\sqrt{121}$ 24. $\sqrt{9}$ 25. $\sqrt{144}$

26. $\sqrt{400}$ 27. $\sqrt{625}$ 28. $\sqrt{256}$ 29. $\sqrt{441}$ 30. $\sqrt{225}$

THINK ABOUT IT!

31. Numbers, for example 1, 8, and 27, are called *perfect cubes* because $1^3 = 1$, $2^3 = 8$, and $3^3 = 27$. Find a number that is both a perfect square and perfect cube.

Measurement: Pythagorean Theorem Name ____________________

Part I: Use the *Pythagorean Theorem* to find the length of each hypotenuse given the lengths of the legs. Round answers to the nearest tenth.

5 ft, 12 ft.

$5^2 + 12^2 = c^2$
$25 + 144 = c^2$
$169 = c^2$
$\sqrt{169} = c$
$\boxed{13 = c}$

The hypotenuse is 13 ft long.

The *Pythagorean Theorem* states:
In a right triangle, the square of the measure of the hypotenuse (longest side) is equal to the sum of the squares of the measure of the legs.

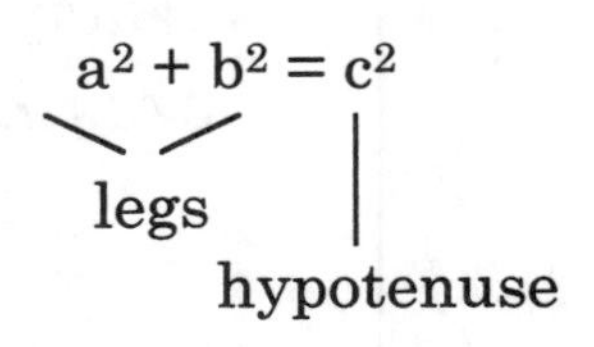

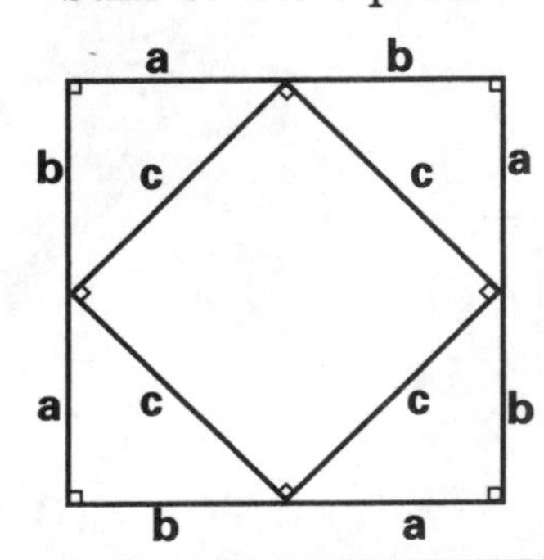

In the figure shown at left, the area of the larger rectangle is equal to the inside square, plus the 4 surrounding triangles. This is represented mathematically in the following equations:

$(a + b)(a + b) = C^2 + 4(\frac{1}{2}ab)$

$a^2 + 2ab + b^2 = C^2 + 2ab$

$a^2 + b^2 = C^2$

---

1. 9 m, 7 m
2. 6 cm, 8 cm
3. 7 in, 24 in
4. 4 ft, 3 ft
5. 10 m, 23 m
6. 2 in, 5 in
7. 3 ft, 7 ft
8. 9 cm, 40 cm
9. 14 ft, 8 ft
10. 9 mm, 13 mm
11. 11 in, 17 in
12. 8 ft, 15 ft

Part II: Given the following lengths, determine whether each triangle is a right triangle. Write *yes* or *no.*

---

13. 7 ft, 24 ft, 25 ft
14. 8 in, 13 in, 16 in
15. 8 cm, 11 cm, 19 cm
16. 9 in, 40 in, 41 in
17. 30 yd, 40 yd, 50 yd
18. 19 m, 20 m, 21 m
19. 9 ft, 12 ft, 15 ft
20. 20 m, 25 m, 30 m
21. 2 in, 3 in, 4 in

Name ____________________

Measurement: Area of Triangles and Trapezoids

Part I: Find the area of each triangle.

Area: measure of what is inside

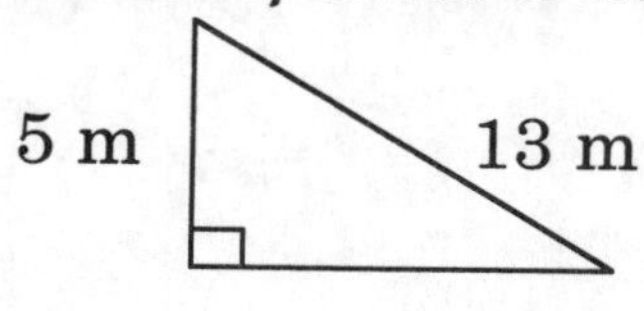

$A = \frac{1}{2}(b \times h)$
$= \frac{1}{2}(5 \times 12)$
$= 30\ m^2$

TIP: *The base and height of a triangle will always be perpendicular.*

1.
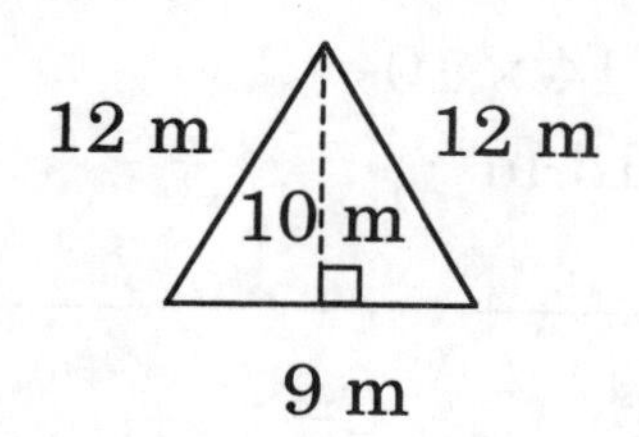

2.
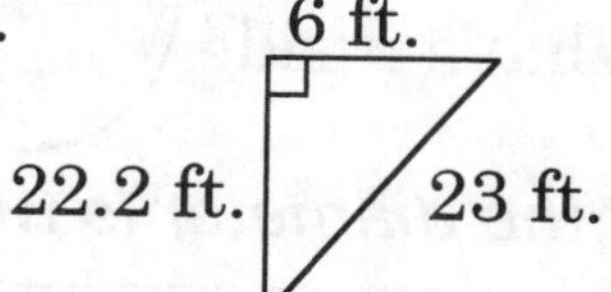

3.
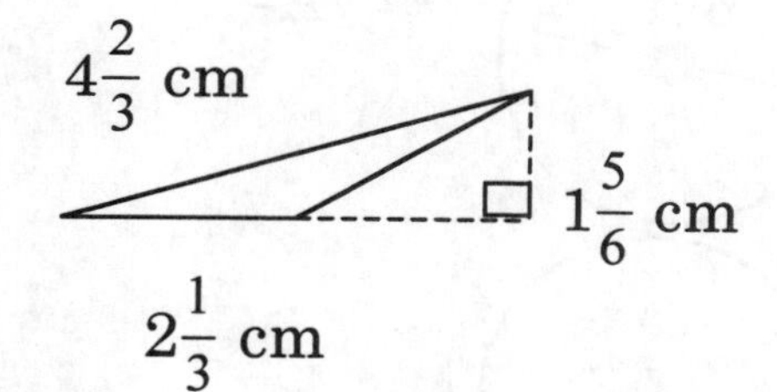

4.
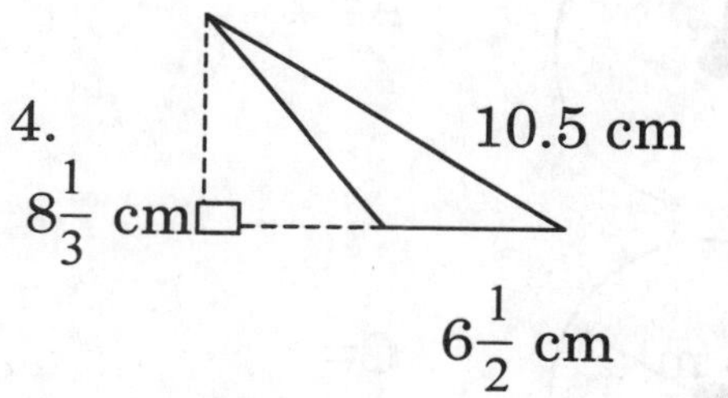

Part II: Find the area of each trapezoid.

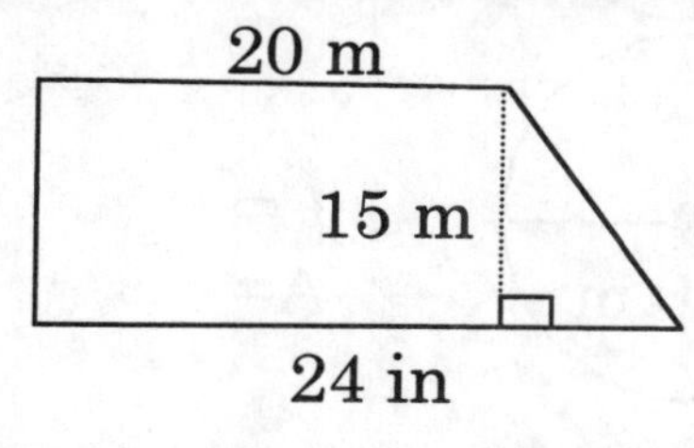

$A = \frac{1}{2}h(a + b)$, where h is the height and a and b are the bases.

$\frac{1}{2}(15)(20 + 24) = 330\ m^2$

5.
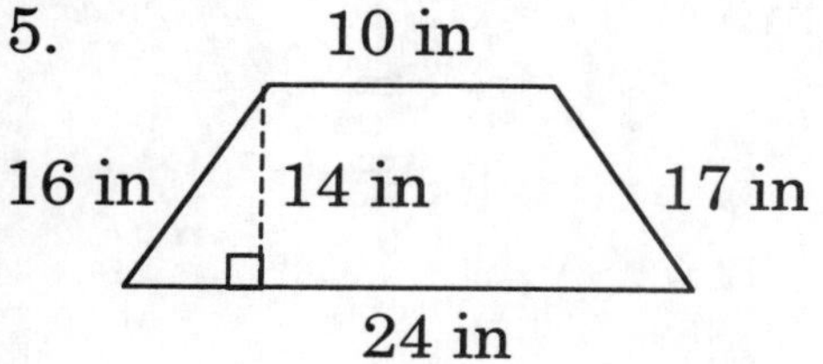

6.
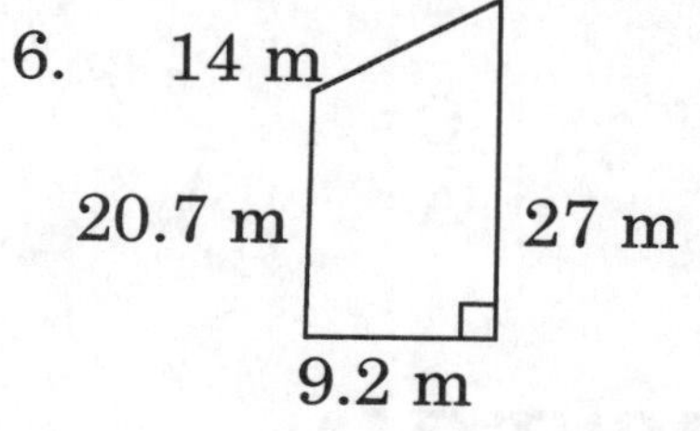

7. bases: 12 m, 18 m
height: 10 m

8. bases: $6^1/_2$ ft, $11^2/_3$ ft
height: 14 ft

Name ____________________

Measurement: Circumference & Area of Circles

Find the circumference and area of each circle. Use 3.14 for π.
Round to the nearest hundredth.

Circumference: Distance around

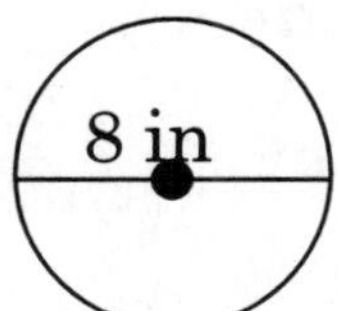

$C = \pi \times d$
$= 3.14 \times 8$
$= 25.12$ in

Area: What's inside

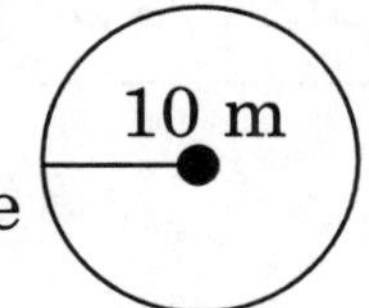

$A = \pi \times r^2$
$= 3.14 \times 10^2$
$= 314\ m^2$

TIP: *Remember that the diameter is twice the radius. d = 2r.*

1.

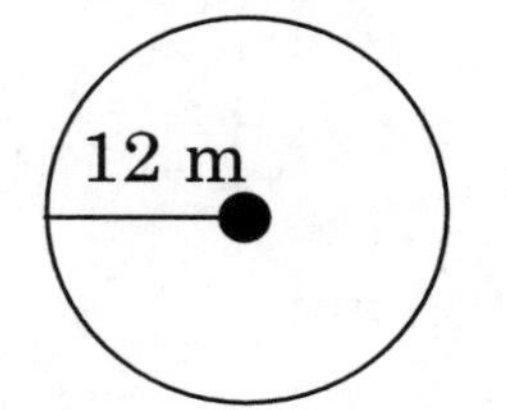

C=
A=

2.

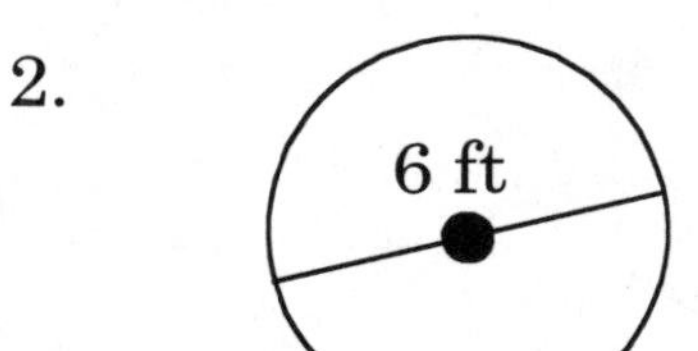

C=
A=

3.

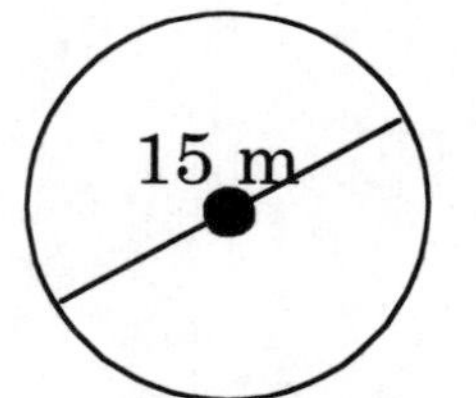

C=
A=

4.

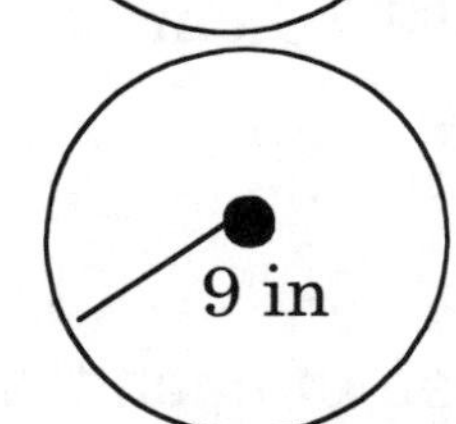

C=
A=

5.

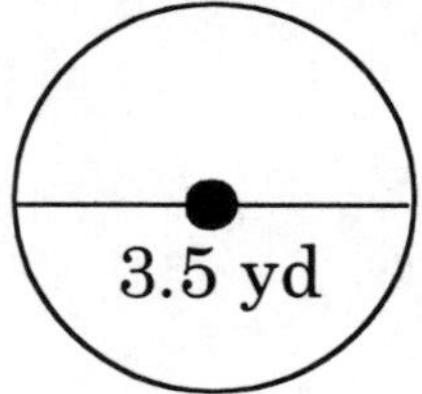

C=
A=

6.

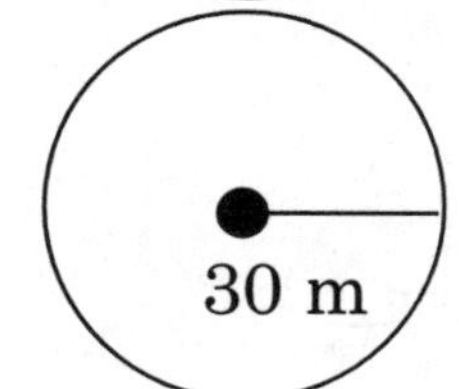

C=
A=

7. d = 7.2 yd

C=
A=

8. r = 7 in

C=
A=

THINK ABOUT IT!

9. A circle has a circumference of 100 in. What is its area?

Measurement: Problem Solving

Name ______________________

Solve each problem.

1. An olympic-sized swimming pool is 50 m long. If Cheryl swims 1 km, how many laps does she swim?

2. A customer orders 1,500 pounds of steel. If the delivery truck holds $^{3}/_{4}$ ton, will the truck be able to deliver the order with one load?

3. What is the length of a rectangle with an area of $31^{2}/_{3}$ square yards and a width of 5 yards?

4. If you cut a cardboard tube and laid it out flat, you will find that it is a rectangle. How much cardboard is used in a tube if its length is $4^{1}/_{2}$ inches and its circumference is 5 inches?

5. Mrs. Hamilton owns a square plot of land that has an area of 10 square miles. Estimate to the nearest whole mile the length of a side of her plot of land.

6. A helicopter rises vertically 800 feet and then travels east 1100 feet. How far is it from its starting point? Round your answer to the nearest whole number.

7. Find the length of the diameter of a circle whose area is 134 m^2. Use 3.14 for π. Round your answer to the nearest tenth.

8. A quarter has a radius of 12 millimeters. Find the area of a quarter.

1.
2.
3.
4.
5.
6.
7.
8.

Name ______________________

Ratio and Proportion: Ratios & Equal Ratios

Part I: Express each ratio as a fraction in simplest form.

9 to 12

$\frac{9}{12} = \frac{3}{4}$

A ratio is a comparison of two numbers. It is often written as fraction in simplest form.

---

1. 15 to 27
2. 21 : 28
3. 14 to 49
4. 125 to 25
5. 11 out of 33
6. 18 : 64
7. 48 hours : 21 hours
8. 21 to 66
9. 2 ft to 6 ft

Part II: Determine whether each pair of ratios is equal. Write ***yes*** or ***no***.

15 16

$\frac{5}{8}, \frac{2}{3}$

no

1. Cross multiply.
2. If the cross products are equal, the ratios are equal.

---

10. $\frac{8}{12}, \frac{24}{32}$
11. $\frac{2}{5}, \frac{10}{25}$
12. $\frac{6}{9}, \frac{9}{12}$
13. $\frac{2}{11}, \frac{6}{33}$
14. $\frac{3}{8}, \frac{12}{32}$
15. $\frac{5}{9}, \frac{8}{12}$
16. $\frac{7}{5}, \frac{8}{6}$
17. $\frac{8}{20}, \frac{6}{15}$
18. $\frac{12}{30}, \frac{3}{8}$
19. $\frac{6}{8}, \frac{3}{4}$
20. $\frac{4}{5}, \frac{28}{35}$
21. $\frac{75}{120}, \frac{40}{90}$

Ratio & Proportion: Rates

Name ______________________

Express each rate as a unit rate.

Rate: a ratio of two measurements with different units
Unit Rate: a rate in which the denominator is 1

\$1.39 per 100 paper plates $\quad$ \$1.39 ÷ 100 = \$0.0139 per 1 plate

---

1. \$2.43 for 3 pounds
2. 220 miles in 5 hours
3. 6 cups for 2 pounds
4. \$12.90 for 10 disks
5. \$450 for 5 days
6. 12 people in 4 cars
7. 9 pounds in 3 weeks
8. 130 tickets in 5 days
9. 20 people in 4 rows
10. 420 miles in 7 days
11. \$29.00 for 4 tapes
12. 2,000 tickets in 4 days
13. 12 cups for 24 pounds
14. \$960 for 16 days
15. 315 miles in 7 hours
16. \$36.00 for 6 tapes

THINK ABOUT IT!

17. The Corner Market sells napkins at \$2.29 for 300 and cups at \$1.75 per 50. Ed's Supermarket sells the same napkins at \$1.49 for 200 and cups at \$0.89 for 25.
    a. Find the unit price for each item at each store.
    b. Which store has the better buy for napkins and for cups?

Ratio & Proportion: Solving Proportions **Name** ____________________

Solve each proportion.

$$\frac{5}{6} = \frac{n}{24}$$

$$6 \bullet n = 5 \bullet 24$$
$$n = 120 \div 6$$
$$n = 20$$

A proportion is two equal ratios. To solve:

1. Find the cross products.
2. Division undoes multiplication to solve for n.

---

1. $\frac{18}{27} = \frac{n}{3}$
2. $\frac{2}{6} = \frac{5}{n}$
3. $\frac{1}{2} = \frac{n}{30}$
4. $\frac{15}{24} = \frac{n}{36}$
5. $\frac{6}{5} = \frac{n}{4}$
6. $\frac{2.5}{4} = \frac{10}{x}$
7. $\frac{3}{4} = \frac{9}{n}$
8. $\frac{8}{12} = \frac{n}{3}$
9. $\frac{15}{9} = \frac{10}{n}$
10. $\frac{5}{n} = \frac{6}{3}$
11. $\frac{18}{42} = \frac{n}{7}$
12. $\frac{2.6}{13} = \frac{8}{h}$
13. $\frac{5}{9} = \frac{n}{5.4}$
14. $\frac{21}{m} = \frac{10}{20}$
15. $\frac{6}{x} = \frac{3}{10}$
16. $\frac{7}{8} = \frac{49}{x}$
17. $\frac{7}{5} = \frac{n}{60}$
18. $\frac{p}{20} = \frac{120}{150}$
19. $\frac{15}{w} = \frac{60}{4}$
20. $\frac{40}{8} = \frac{150}{c}$
21. $\frac{4}{3} = \frac{n}{45}$
22. $\frac{7}{8} = \frac{n}{72}$
23. $\frac{n}{72} = \frac{2}{4}$
24. $\frac{50}{60} = \frac{n}{3}$

THINK ABOUT IT!

25. Use the clues to write a proportion.
    Each ratio is equal to 2/3. The numerator for one ratio is 8 and the denominator for the other ratio is 18.

Ratio & proportion: Problem Solving

Name ____________________

Solve each problem.

1. Use the rectangles below to complete the following:
   a. Write ratios comparing the widths, the lengths, and the perimeters of rectangle y to z.
   b. Show whether or not these ratios are equivalent.

   Rectangle y: 6 in by 9 in; Rectangle z: 8 in by 12 in

2. Caroline found that 6 students out of 18 students she surveyed liked sushi. Josh found that 9 students out of 24 students he surveyed liked sushi. Which result shows a greater preference for sushi?

3. The band has 5 days to sell 195 tickets to ensure a sellout at their fall concert. At what rate must they sell the tickets?

4. If a race car uses 323 liters of gasoline in a 500 kilometer race, about how many liters were used for each kilometer?

5. A supermarket has soda on sale, 6 cans for $1.95. Each can sold separately costs $0.35. How much do you save buying the 6 cans on sale?

6. Creek Middle School has 1,000 students, 40 teachers, and 5 administrators. If the school grows to 1,200 students and the ratios are maintained, find the number of teachers and administrators that will be needed.

Answers
1.
2.
3.
4.
5.
6.

Percent: What is a Percent?

Name ____________________

Part I: Write each ratio as a percent.

52:100 = 52%

$\frac{74}{100} = 74\%$

A percent is a special ratio that compares a quantity to 100.

---

1. 91:100
2. $\frac{23}{100}$
3. $\frac{17}{100}$
4. 28:100
5. 47 to 100
6. 18:100
7. 15 to 100
8. $\frac{63}{100}$
9. 44 to 100
10. 33:100
11. $\frac{21}{100}$
12. 25:100

Part II: Write each percent as a ratio (use fractions in lowest terms).

13. 70%
14. 95%
15. 45%
16. 86%
17. 33%
18. 84%
19. 50%
20. 20%
21. 15%
22. 75%
23. 25%
24. 10%

THINK ABOUT IT!

25. Use each diagram below to show 60%.

a. 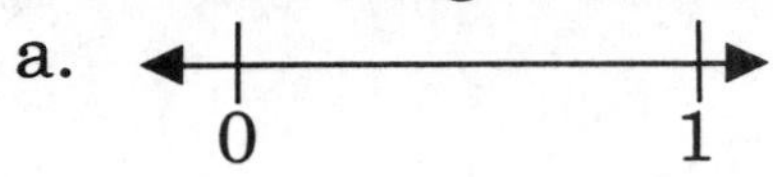

b. 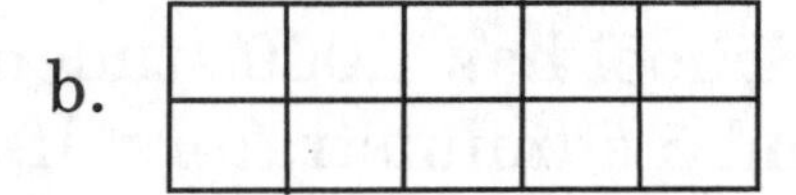

c.

d.

Percents: Percents & Fractions

Name ______________________

Part I: Express each fraction as a percent. Round to the nearest hundredth.

$\frac{12}{20}$

$12 \div 20 = .6$

60%

1. Change the fraction to a decimal. (numerator ÷ denominator)
2. Change the decimal to a percent. (Move the decimal two places to the right. Add a zero if necessary, and do not forget the percent sign.)

TIP: *A percent compares a quantity to 100. If the fraction is out of 100, then your numerator is the percent.* $\frac{43}{100}$ *= 43%*

---

1. $\frac{3}{8}$ 2. $\frac{24}{25}$ 3. $\frac{18}{25}$ 4. $\frac{2}{5}$

5. $\frac{3}{10}$ 6. $\frac{11}{20}$ 7. $\frac{2}{3}$ 8. $\frac{3}{3}$

9. $\frac{40}{125}$ 10. $\frac{5}{6}$ 11. $\frac{1}{4}$ 12. $\frac{1}{8}$

Part II: Express each percent as a fraction in lowest terms.

$55\% = \frac{55}{100} = \frac{11}{20}$

1. Drop the percent sign and place the number over 100.
2. Reduce to lowest terms.

---

13. 15%   14. 25%   15. 72%   16. 10%

17. 70%   18. 50%   19. 34%   20. 2%

21. 42%   22. $3\frac{1}{3}\%$ 23. 1%   24. 23%

Percent: Percents & Decimals

Name ____________________

Part I: Express each decimal as a percent.

0.4
$0.4 \times 100 =$
40%

To change a decimal to a percent, multiply by 100 and add the percent sign.

TIP: *When you multiply by 100, it moves the decimal point two places to the right. Simply moving the decimal point two places to the right changes the decimal to a percent.*

---

1. 0.39
2. 0.75
3. 0.875
4. 0.1
5. 0.6
6. 0.06
7. 0.045
8. 0.999
9. 0.46
10. 0.05
11. 0.565
12. 0.5

Part II: Express each percent as a decimal.

54%
.54

To change a percent to a decimal, move the decimal point two places to the left, and take away the percent sign.
3% = .03   40% = .4

---

13. 43%
14. 89%
15. 7%
16. 2%
17. 6.2%
18. 34.5%
19. 90%
20. 17%
21. 100%
22. 51%
23. 310%
24. 79%

Percents: Large & Small Percents

Name ____________________

Part I: Express each percent as a decimal.

400% = 4 — To change a percent to a decimal, move the decimal point two places to the left.

1. 0.032%
2. 340%
3. 0.014%
4. 0.0035%
5. 100%
6. 0.078%
7. 425%
8. $\frac{1}{5}$%

Part II: Express each number as a percent.

$6\frac{1}{2} = 6.5 = 650\%$ — $0.008 = 0.008 = 0.8\%$

9. 5
10. 70
11. 2.26
12. 0.03
13. $3\frac{1}{4}$
14. $9\frac{1}{2}$
15. $5\frac{9}{10}$
16. $7\frac{2}{5}$
17. 385
18. 2.7
19. 0.0035
20. 4
21. $4\frac{3}{4}$
22. $8\frac{1}{5}$
23. $10\frac{1}{2}$
24. $12\frac{1}{4}$

THINK ABOUT IT!

25. Replace each ____ with <, >, or = .

a. 1.5 _____ 150%   b. 560 _____ 5,600%   c. 14 _____ 14,000%

Percent: Percent of a Number

Name ______________________

Use a proportion to solve each problem. Round answers to the nearest tenth.

Percent Proportion

$$\frac{\text{Part}}{\text{Whole}} = \frac{\%}{100}$$

What number is 25% of 520?

$$\frac{x}{520} = \frac{25}{100}$$

$$100x = 13{,}000$$

$$x = 130$$

Answer: **130** is 25% of 520.

1. Identify the part, whole, and/or percent.
2. Plug the numbers into the proportion and solve for the missing piece (part, whole or %).

---

1. 25 is what % of 40?
2. Find $37\frac{1}{2}$% of 64.
3. 50% of 128 is what number?
4. What number is 60% of 72?
5. 2 is 40% of what number?
6. Find 80% of $90\frac{1}{2}$.
7. 107 is what percent of 214?
8. What number is 25% of 36?
9. Find 92% of 120.
10. 40% of 80 is what number?
11. 55 is what percent of 60?
12. 12% of 16.5 is what number?
13. What number is 20% of 20?
14. 3.9 is what percent of 14?

Percent: Percent of Change

Name ______________________

Find the percent of increase or decrease. Round to the nearest whole percent.

old: 8 (original)
new: 15
$15 - 8 = 7$ ← 1. Find the amount of increase or decrease.

$\frac{7}{8} = \frac{x}{100}$ ← 2. Write a proportion: $\frac{\text{change}}{\text{original}} = \frac{x}{100}$

$700 = 8x$
$x = 88\%$ ← 3. Solve the proportion to find the percent of change.

---

1. old: $4
   new: $7

2. old: 36
   new: 18

3. old: 30
   new: 24

4. old: 35
   new: 70

5. old: $75
   new: $60

6. old: $6.80
   new: $8.20

7. old: $456
   new: $500

8. old: 60
   new: 38

9. old: 40
   new: 80

10. old: $126
    new: $150

11. old: 275
    new: 150

12. old: $35
    new: $45

13. old: 62
    new: 50

14. old: 0.95
    new: 1.6

15. old: 9.8
    new: 10

THINK ABOUT IT!

16. A book is on sale for $14, which is 30% off the original price. Can you find the original price by adding 30% of $14 to the sale price? Explain.

Percent: Discount & Sales Tax

Name ______________________

Find the discount or sales tax to the nearest cent.

$400 CD player Discount: 30% What is 30% of $400? .3 x 400 = 120 **$120 is the discount.**	$75 sports jacket Sales Tax: 6% What is 6% of $75? .06 x 75 = 4.5 **The sales tax is $4.50.**

1. $15 CD
   25% discount

2. $36 sunglasses
   20% discount

3. $49 shoes
   7% sales tax

4. $15.99 t-shirt
   20% off

5. $175.95 suit
   6% sales tax

6. $25 watch
   30% discount

7. $9.95 cassette
   8% sales tax

8. $20 book
   7% sales tax

9. $21.95 snow shovel
   60% discount

10. $2.50 socks
    15% discount

THINK ABOUT IT!

11. A sweatshirt regularly sells for $38. It is on sale at a 15% discount with 6% sales tax.
    a. Does it matter in which order the discount and the sales tax are applied? Explain.
    b. Would the final price change if the sales tax were added before the discount was subtracted? Explain.

Percent: Problem Solving

Name ____________________

Solve each problem.

1. Mrs. Truett bought a dining room set that cost $1,200. The store required a 30% down payment to hold the furniture. How much was the down payment?	1.
2. The ski club is going on a skiing trip to Colorado. There are 52 members. If 75% of the members sign up for the trip, how many are going skiing in Colorado?	2.
3. Twenty-eight of the 131 students in Ms. Martin's math classes received A's on the last test. *About* what percent of the class earned A's?	3.
4. Jerry answered 28 of the 35 questions correctly on his history test. What percent of the questions did Jerry answer correctly?	4.
5. A class pool party was attended by 80% of the students. If there are 60 students in the class, how many attended the party?	5.
6. Jeanette was once able to do 10 push-ups. Now she can do 25. Find the percent of increase.	6.
7. Ms. Miller bought a new suit that cost $175. She bought it on sale, though, for 40% off. a. What was her discounted amount? b. How much did she pay for the suit on sale?	7.

Name ____________________

Patterns & Functions: Arithmetic & Geometric Sequences

Find the next three terms in each sequence.

Arithmetic Sequence
The difference between each pair of numbers is the same.

3, 6, 9, 12, . . . 15, 18, 21
+3

Geometric Sequence
You can always find the next term by multiplying the previous term by the same number.
3, 9, 27, 81, . . .243, 729, 2187
× 3

---

1. 7, 14, 21, 28 . . .
2. 1, 4, 16, 64, . . .
3. 5, 10, 20, 40, . . .
4. 15, 30, 45, 60, . . .
5. 9, 3, 1, $\frac{1}{3}$, $\frac{1}{9}$, . . .
6. $\frac{1}{16}$, $\frac{1}{4}$, 1, 4, . . .
7. 1, 2.1, 3.2, 4.3, . . .
8. 0, 17, 34, 51, . . .
9. 2, 1, 0.5, 0.25, . . .
10. 6, 12, 18, 24, . . .
11. 1, 5, 25, 125, . . .
12. 0.2, 0.4, 0.8, 1.6, . . .
13. 0, 16, 32, 48, . . .
14. 64, 32, 16, 8, . . .

THINK ABOUT IT!

15. Create a sequence using the given rules. Provide at least four terms for each sequence beginning with the given number.
    a. Add 0.6 to each term: 10
    b. Multiply each term by $\frac{1}{2}$: 4

MATH FACTS

Modern mathematics has been built throughout the past 1,800 years by contributions from China, India, the Arabic Empire, East & West Europe, & America

Patterns & Functions: Problem Solving Name ____________

Solve each problem.

1. Brenda decides to start exercising. She knows that she must begin gradually. She decides to work out 5 minutes the first day, and then double her exercise time each day for a week. Write a sequence showing the length of time she will exercise each day. Is her plan reasonable? Why or why not?

2. A parking garage in Atlanta charges $2 for the first hour then $0.75 for each additional hour. Stacey parks her car in the garage at 8:00 a.m. and owes $8.75 when she leaves. How long was her car in the garage?

3. Sue baked some chocolate chip cookies. She kept half of them for herself and took the other half to work. She put the cookies in 4 bags, a dozen in each bag. How many cookies did she make?

4. Mr. Owens is delivering bottles of soda to food marts. At the first one, he drops off half of the bottles in the truck. At each of the other marts, he delivers half of the bottles he has left in the truck. At the eleventh store, he drops off 1 bottle, which is the last one. How many bottles were originally in the truck?

1.
2.
3.
4.

Probability: Tree Diagrams **Name** ____________________

Make a tree diagram to show all the outcomes for each situation.

A tree diagram is used to show the total number of possible outcomes in a probability experiment.

Flipping a quarter and a nickel.

```
      ,H    HH
    H<
      `T    HT
      ,H    TH
    T<
      `T    TT
```

There are 4 possible outcomes.

1. ordering a hamburger, hot dog, or chicken and coke, diet coke, or sprite

2. choosing a small, medium, or large shirt in red or blue

3. choosing a number 1, 2, or 3 and choosing a letter x, y, or z

4. choosing pepperoni or ham pizza and thick or thin crust

5. choosing a white, black, or red car with 2 or 4 doors

Name ______________________

Probability: Fundamental Counting Principle

In each situation, find the total number of outcomes.

Sam has to wear a uniform to school. He can choose from 5 different kinds of shirts and 4 different kinds of pants/shorts. How many different combinations of tops and bottoms are there?

Multiply the number of choices in each set to derive the number of possible combinations.

5 shirts × 4 pants/shorts = 20 combinations

---

1. selecting 1 shirt from 12 different shirts and 1 tie from 14 different ties

2. choosing an interior color and exterior color for a new van if there are 5 interior colors to choose from and 8 exterior colors

3. choosing the first two digits for a telephone number if you can use the digits 1-9 for the first number and 0-9 for the second

4. choosing from 8 different kinds of sandwiches for lunch and 6 different kinds of side dishes

5. flipping a penny, nickel, dime and quarter

6. choosing from one of two English courses, four mathematics courses, three science courses, and five social studies courses

7. choosing from three different sweaters, four different slacks, and three different pairs of socks

Name ______________________

Probability: Experimental & Theoretical Probability

Read each problem. Then solve.

Experimental probability	***Theoretical Probability***
The probability based on the outcomes of an experiment.	The probability based on mathematical principles.

Example: Find the theoretical probability of a coin landing heads up.

$$P\,(H) = \frac{1}{2}$$

1 ⟵ number of heads
2 ⟵ number of possible outcomes

---

1. Find the theoretical probability of choosing a girl's name from 20 girls' names and 10 boys' names.

2. Jennifer tosses a coin 20 times. It lands on tails 12 times.
   a. What is the experimental probability of getting tails?
   b. How does the theoretical probability of getting tails compare with the experimental probability?

3. Leland tosses two coins four times, and twice both coins landed on tails.
   a. What is the experimental probability of getting two tails?
   b. What is the theoretical probability of getting two tails?
   c. What is the theoretical probability of getting two heads?
   d. What is the theoretical probability of getting a head and a tail?

4. Mrs. Mendel has a bag with two blue chips, two red chips, and three white chips.
   a. What is the theoretical probability of choosing a red or blue chip?
   b. What is the theoretical probability of not choosing a blue chip?

Probability: Problem Solving

Name ____________________

Solve each problem.

1. The combination to your locker has a three-digit combination. Each digit can be a number from one to nine. How many different combinations are possible?

2. The football team is choosing a uniform. They can choose from blue or white socks, black or white shoes, 3 different colors of pants, and 8 different colors of jerseys. How many different clothing combinations do they have?

3. Three out of ten students at Cory's school ride a bike to school. If there are 850 students at his school, how many ride bikes to school?

4. Mrs. Miller has a container of jellybeans on her desk. There are 300 jellybeans in the jar. 65 are cherry-flavored, 70 are apple-flavored, 85 are blueberry-flavored, 45 are coffee-flavored, and 35 are popcorn-flavored. What is the probability of picking the given flavor(s).
   a. cherry
   b. blueberry or apple
   c. any flavor but coffee
   d. popcorn

5. Steve is making a sandwich. He has two kinds of bread, three kinds of meat, and four kinds of condiments. If he uses only one of each, how many different sandwiches can Steve make?

1.
2.
3.
4.
5.

Answer Key

Diagnostic Test

Page 5 1. B, 2. D, 3. C, 4. B, 5. B, 6. B, 7. B, 8. B, 9. C, 10. B, 11. C, 12. B, 13. B, 14. C, 15. D, 16. B, 17. D, 18. B

Page 6 19. C, 20. D, 21. C, 22. A, 23. C, 24. B, 25. A, 26. D, 27. A, 28. C, 29. A

Page 7 30. C, 31. C, 32. B, 33. A, 34. C, 35. C, 36. B, 37. C, 38. B, 39. A, 40. B

Page 8 41. C, 42. B, 43. B, 44. C, 45. C, 46. C, 47. C, 48. B, 49. C, 50. C, 51. A, 52. B, 53. B

Page 9 1. 1,000, 2. 8, 3. 70, 4. 4,000, 5. 30, 6. 200, 7. 3,600, 8. 30, 9. 2, 10. 40, 11. 3,800, 12. 5, 13. 80, 14. 50, 15. 820, 16. 80, 17. 300, 18. 300, 19. 800, 20. 10,000, 21. 9,000, 22. 100, 23. 7,000, 24. 2,000

Page 10 1. 36,315, 2. 51,625, 3. 43,500, 4. 49,898, 5. 88,321, 6. 35,099, 7. 130,247, 8. 5,749, 9. 110,192, 10. 33,689, 11. 50,987, 12. 36,809, 13. 39,159, 14. 70,919, 15. 35, 16. 61,443, 17. 216,478, 18. 112,515, 19. 193,432, 20. 228,756, 21. 11,377, 22. 21,046, 23. 19,631

Page 11 1. 47,100, 2. 2,882,000, 3. 9,833 R3, 4. 477,688, 5. 134 R6, 6. 387,450, 7. 69 R10, 8. 975,768 9. 924, 10. 1,155,276, 11. 804 R7, 12. 11 R 706

Page 12 1. 7•7•7•7•7, 2. 2•2•2•2•2•2, 3. 3•3•3•3•3 4. 5•5•5, 5. 4•4•4•4, 6. 12•12, 7. 10•10•10•10•10 8. 5•5•5•5, 9. 4•4•4, 10. 2^3, 11. 13^2, 12. $5^2 \cdot 6^2$, 13. $4 \cdot 5^2$, 14. $7^3 \cdot 8^2$, 15. $4^2 \cdot 8^2 \cdot 9$, 16. $2^2 \cdot 4^3$, 17. $9^2 \cdot 10$, 18. $10^2 \cdot 11^2$, 19. 121, 20. 12,500, 21. 2,410, 22. 6,912, 23. 316, 24. 1,417,176

Page 13 1. 36, 2. 8, 3. 2, 4. 142, 5. 12, 6. 20, 7. 31, 8. 19, 9. 49, 10. 16, 11. 4, 12. 21, 13. 23, 14. 14, 15. 0, 16. +, +, ×

Page 14 1. 63, 2. $315, $16,380, 3. $75, 4. 10 minutes, 5. $12, $8, 6. 21,200 pounds, 7. $23

Page 15 1. >, 2. <, 3. <, 4. <, 5. <, 6. <, 7. <, 8. =, 9. <, 10. >, 11. <, 12. <, 13. <, 14. >, 15. >, 16. <, 17. Possible Answer: 5.295

Page 16 1. 3.45427, 3.45572, 3.45725, 2. 224.59993, 224.90553, 224.95003, 3. 0.055753, 0.055755, 0.07505, 4. 22.21212, 22.21221, 22.22111, 5. 4.49944, 4.499441, 4.49994, 6. 0.73323, 0.73342, 0.73422, 7. 3,617.03, 3,671.03, 3,671.031, 8. .456, 400.56, 456, 9. 314.00991, 314.09091, 314.09901, 10. 0.08183, 0.08813, 0.08883, 11. 79.06542, 79.60006, 79.60054 12. 5.00387, 5.03008, 5.037708, 13. 0.0012002, 0.001202, 0.001210, 14. 0.443344, 0.443434, 0.443443

Page 17 1. 10, 2. 80, 3. 100, 4. 60, 5. 15, 6. 50, 7. 600, 8. 30, 9. 40, 10. 150, 11. 200, 12. 400, 13. 16, 14. 210, 15. 400, 16. 10, 17. 10, 18. 18, 19. 1,500, 20. 10, 21. 24, 22. 30, 23. 600, 24. 20, 25. $2.00

Page 18 1. 370.73, 2. 10.11101, 3. 10.448, 4. 1.167791, 5. 190.98444, 6. 24.8239, 7. .46678, 8. 11.9977, 9. 89.4978, 10. 6.366, 11. .3369, 12. .7, 13. 1.025, 14. 8.62, 15. .751, 16. 3.256, 17. 6.588, 18. 1.622, 19. 2.73, 20. 19.44, 21. 74.83

Page 19 1. 100, 2. 81, 3. 7, 4. 300, 5. 8, 6. 40, 7. 4, 8. 64, 9. 9, 10. 400, 11. 2, 12. 2,400, 13. 3, 14. 70, 15. 6, 16. 5, 17. 7, 18. 3, 19. 200, 20. 4, 21. 900, 22. 7, 23. 63, 24. 100, 25. 6, 26. 6, 27. 30

Page 20 1. .48, 2. .243, 3. 12.261, 4. 38.88, 5. .0252, 6. 7.868, 7. 9.063, 8. .70875, 9. 205, 10. .68, 11. 87, 12. .00115, 13. 651, 14. 10.02, 15. 20, 16. 3.6

Page 21 1. 8.2×10^2 2. 8.4×10^{-3} 3. 2.4×10^3 4. 6.324×10^3 5. 3×10^{-5}, 6. 6.784×10^6 7. 8.6×10^4 8. 9.3×10^{-4} 9. 5.1×10^4 10. 220,000 11. 6,104 12. .00125 13. 900,000,000 14. .00000917 15. .000196 16. .000067 17. 29,830 18. 74 19. 3.2×10^{-3}, 9.7×10^6, 5.1×10^7

Page 22 1. 7 posters, 2. about 5 miles, 3. \$97.75, 4. \$7.04; \$12.96, 5. \$79.60, 6. 9 hours

Page 23 1. t + 7, 2. r + 2, 3. p – 8, 4. g – 4, 5. c ÷ 7, 6. 3a, 7. $\frac{b}{2}$, 8. 7n, 9. b – 6, 10. 9 + c, 11. 35, 12. 6, 13. 17, 14. 150, 15. 7, 16. 250, 17. 2,500, 18. 22, 19. 2

Page 24 1. n + 14 = 25, 2. n – 12 = 25, 3. 3 + 5y = 33, 4. 5c + 3 = 18, 5. $\frac{n}{6} = 5$, 6. 6 + 7n = 75, 7. s + .10 = 2.09, 8. 7x – 5 = 58, 9. $8 + \frac{c}{7} = 11$, 10. 3w + 12 = 450, 11. 2d – 14 = 48, 12. $\frac{c}{5} - 4 = 10$, 13. 2s + 1 = 53, 14. 51 = 4b – 13, 15. Possible Answer: 6 times a number, less 2, is 3.

Page 25 1. a. 5< n, b. n – 5, c. 5 – n, 2. a. s + 100, b. s – 50, c. 3s, 3. Possible Answers: a. 12 times d; the product of 12 and d, b. 850 less n; 850 minus n, c. x divided by 5; the quotient of x and 5, d. the sum of y and 14; y plus 14, 4. 55¢, 5. a. 0, b. no; ab will always equal ba because of the commutative property of multiplication so their difference will always be zero.

Page 26 1. d = 38, 2. x = 48, 3. t = 61, 4. r = 1.5, 5. g = 4.75, 6. n = 60, 7. x = .7, 8. n = 530, 9. n = 408, 10. y = 4.09, 11. b = 1.7, 12. s = 9.2, 13. x = 2.3, 14. x = 82, 15. x = 4.4, 16. n = 45, 17. x = 265, 18. m = 234, 19. x = 317, 20. x = 64, 21. c = 45, 22. r = 12, 23. x = .8, 24. r = 10.2

Page 27 1. n = 16, 2. x = 4, 3. n = 8, 4. n = 113.4, 5. x = 45, 6. n = 270, 7. n = 1.4, 8. n = 172.8, 9. n = 136, 10. y = 14, 11. x = 42, 12. y = 158, 13. n = 6.8, 14. c = 32, 15. y = 20, 16. n = 7.5, 17. h = 4.5, 18. x = 36, 19. n = 144, 20. x = 140, 21. s = 20, 22. r = 85, 23. c = 7, 24. n = 2,546

Page 28 1. x = 31, 2. y = 2.99, 3. n = 65, 4. r = 36, 5. x = 25, 6. c = 4, 7. r = 8, 8. n = 0.2, 9. t = 25.3, 10. x = 12.96, 11. n = 81.5, 12. d = 19, 13. x = 3.7, 14. y = 18, 15. m = 2.5, 16. n = 201, 17. n = 2.3, 18. m = 43, 19. x = 35, 20. y = 3, 21. r = 21, 22. c = 83.7, 23. m = 646, 24. h = 5.4, 25. n = 749.7, 26. n = 15, 27. n = 123, 28. h = 9, 29. n = 2.02, 30. n = 70

Page 29 1. n = 4, 2. x = 20, 3. y = 4, 4. x = 0, 5. n = 15, 6. b = 175, 7. x = 17, 8. x = 100, 9. r = 9, 10. h = 21, 11. n = 10, 12. n = 3, 13. n = 6, 14. r = 4, 15. e = .5, 16. f = 192, 17. y = 3, 18. x = 3, 19. y = 4, 20. x = 6, 21. m = 7

Page 30 1. (n – 11) • 2 = 20; n = 21; 2. $\frac{x}{9} + 3 = 6$; x = 27; 3. 2 + (2b) = 8; b = 3; 4. (x + 5) –1 = 6; x = 2; 5. 5n – 6 = (25 –4) -2; n = 5; 6. $\frac{b}{4} = 3$; b = 12; 7. 15n = 180; n = 12; 8. $\frac{n}{5} + 6 = 11$; n = 25; 9. $\frac{r}{3} = q$; q = 3

Page 31 1. –12, 2. 6, 3. –34, 4. –11, 5. 50, 6. 2, 7. –15, 8. 20, 9. 4, 10. –3, 11. 2000, 12. –100, 13. –7, 14. 16, 15. –3, 16. 6, 17. –210, 18. 12, 19. 100, 20. 25,
21. -6 -4 -2 0 2 4

Page 32 1. 4, 2. 5, 3. 11, 4. 3, 5. 0, 6. 6, 7. 8, 8. 12, 9. 23, 10. 9, 11. 45, 12. 33, 13. 28, 14. 51, 15. 61, 16. 73, 17. 18, 18. 19, 19. 14, 20. 38, 21. 4, 22. 6, 23. 15, 24. 124, 25. A. –10, B. 5, C. –32

Page 33 1. < 2. < 3. > 4. > 5. > 6. = 7. > 8. > 9. > 10. > 11. < 12. < 13. 1, 0, -1, -4, -5, -8, 14. 7, 3, 0, -2, -7 15. 14, 1, 0, -1, -5, -9 16. 110, 60, 0, -20, -60, -140 17. 50, 20, 10, -10, -20, -30 18. 60, 40, 0, -10, -30, -40 19. 90, 85, 50, -25, -75, 20. -5, -10, -17, -18, -20

Page 34 1. 24, 2. 5, 3. 25, 4. –13, 5. –16, 6. 105, 7. –50, 8. –21, 9. –5, 10. –7, 11. 7, 12. 0, 13. –60, 14. –18, 15. 150, 16. –12 , 17. 120, 18. 3, 19. 64, 20. –56, 21. 60, 22. 1, 23. –400, 24. –72

Page 35 1. 4, 2. 30, 3. –12, 4. –14, 5. –8, 6. 14, 7. 34, 8. –3, 9. –14, 10. –55, 11. –26, 12. 59, 13. –68, 14. 70, 15. –81, 16. –108, 17. –132, 18. 0, 19. –1, 20. –2, 21. 20, 22. a. never b. sometimes, c. always

Page 36 1. –3, 2. –8, 3. –10, 4. –7, 5. 13, 6. 4, 7. –27, 8. –15, 9. 16, 10. –18, 11. 3, 12. 11, 13. 78, 14. 91, 15. –76, 16. 135, 17. –78, 18. –72, 19. 0, 20. –90, 21. 50, 22. The answers are opposite; yes.

Page 37 1. x = 16, 2. m = 2, 3. y = $^{-}$19, 4. x = $^{-}$5, 5. x = 46, 6. a = $^{-}$4, 7. a = $^{-}$392, 8. x = $^{-}$72, 9. c = $^{-}$40, 10. y = $^{-}$168, 11. h = $^{-}$480, 12. t = $^{-}$401, 13. x = 92, 14. x = 43, 15. m = $^{-}$56, 16. x = $^{-}$43, 17. p = $^{-}$175, 18. r = 0, 19. x = $^{-}$10, 20. c = $^{-}$5, 21. t = $^{-}$190, 22. d – ($^{-}$4) = 10; d = 6

Page 38 1. z, y, w, x, 2. a. $150 + ($^{-}$$77) = x b. $73, 3. T + 35 seconds or 35 seconds after liftoff, 4. a. –60, b. 60, c. negative, d. If there is an even number of negatives, the product is positive. If there is an odd number of negatives, the product is negative. 5. –2 cm/minute

Page 39 1. black, white, blue, red, 2. no, 3. Possible Answer: Revote with only black and white, 4. 71%; 29%, 5. t-shirt

Page 40 1. B, 2. C, 3. D, 4. B, 5. A, 6. A

Page 41 1. 1,000, 2. 160, 3. 450, 4. 2,500, 5. 900, 6. 100, 7. blue = 125; red = 50; yellow = 75, 8. 15

Page 42 1. 4.4; 4.5; 3 and 5, 2. 5.3; 5.5; 7, 3. 93; 93; 90 and 94, 4. 16; 17; 17, 5. 52.7; 50; 60, 6. 65.5; 65; 65 7. 9.0; 9.1; none, 8. 132.7; 138.5; 144, 9. 1,780; 1,780; 1,805, and 1,755, 10. 292.9; 250; 250

Page 43 1. a. 26.6, 28, 20, b. mode, 2. mode; appears most often, 3. Answers will vary: a. 1, 2, 3, 10, 11, b. 1, 2, 2, 2, 3, c. 4, 5, 7, 8, 9 4. Possible Ans: no; it is not a representative group, 5. Answers may vary: How many people were surveyed or the age group of the people.

Page 44 1. right, 2. straight, 3. acute, 4. obtuse, 5. acute, 6. right, 7. acute, 8. obtuse, 9. straight, 10. obtuse, 11. acute, 12. right, 13. A. b, B. d, C. a, D. e, E. c

Page 45 1. isosceles, obtuse, 2. scalene, right, 3. scalene, obtuse, 4. equilateral, acute, 5. isosceles, right, 6. isosceles, acute, 7. scalene, obtuse, 8. scalene, right, 9. Possible Answers: a. b.

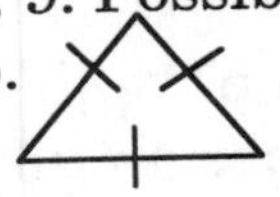

c.

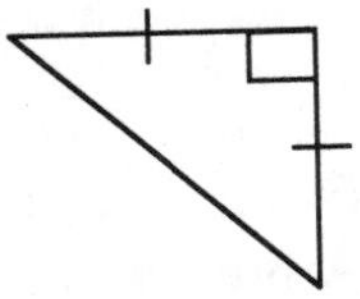

Page 46 1. Rectangle; Parallelogram, 2. Parallelogram, 3. Rectangle; Parallelogram, 4. Trapezoid, 5. Square; Rectangle; Rhombus Parallelogram, 6. Square; Rectangle; Rhombus; Parallelogram, 7. a. square, rectangle, parallelogram, b. yes; opposite sides are equal and all angles equal 90°. c. no, they do not have to have 4 right angles and all sides equal, d. A trapezoid has only 1 pair of parallel sides, but a parallelogram has 2 pairs of parallel sides.

Page 47 1. x = 60°, 2. x = 33°, 3. x = 109°, 4. x = 48°, 5. x = 84°, 6. x = 95°, 7. 66°

Page 48 1. 2. none 3. infinite

4. 5. 6.

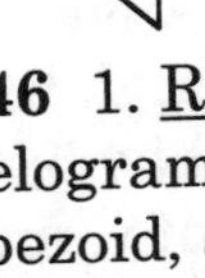

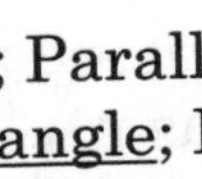

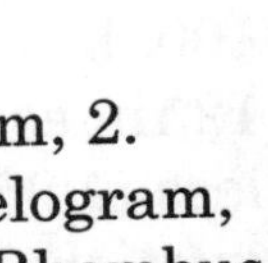

7. no 8. yes 9. yes

Page 49 1. (-8, 8) 2. (2, -2), 3. (-7, -8), 4. (4, 7) 5. (0, 6), 6. (6, 0)

Page 49 continued

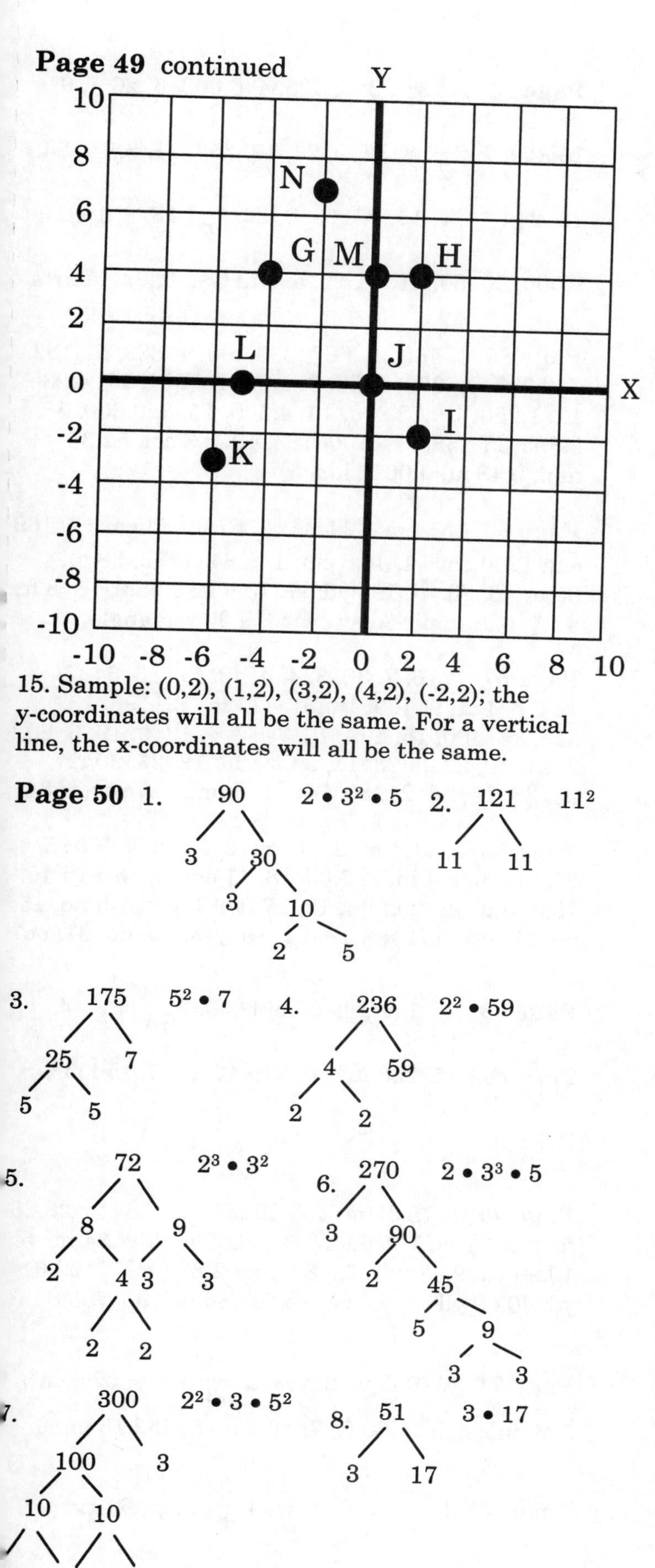

15. Sample: (0,2), (1,2), (3,2), (4,2), (-2,2); the y-coordinates will all be the same. For a vertical line, the x-coordinates will all be the same.

Page 50 1. 90 $2 \bullet 3^2 \bullet 5$ 2. 121 11^2

3. 175 $5^2 \bullet 7$ 4. 236 $2^2 \bullet 59$

5. 72 $2^3 \bullet 3^2$ 6. 270 $2 \bullet 3^3 \bullet 5$

7. 300 $2^2 \bullet 3 \bullet 5^2$ 8. 51 $3 \bullet 17$

Page 51 1. 15, 2. 10, 3. 13, 4. 6, 5. none, 6. 4, 7. 18, 8. 9, 9. 12, 10. 12, 11. 9, 12. 26, 13. 2, 14. 5, 15. 6

Page 52 1. 15, 2. 60, 3. 240, 4. 900, 5. 3,750, 6. 176, 7. 48, 8. 72, 9. 72, 10. 252, 11. 480, 12. 448, 13. 66, 14. 90, 15. 182

Page 53 1. .24, 2. .45, 3. .03, 4. .26, 5. .18, 6. .67, 7. .88, 8. 1.38, 9. .73, 10. .06, 11. 1.12, 12. .58, 13. $\frac{1}{2}$, 14. $\frac{16}{25}$, 15. $\frac{9}{100}$, 16. $\frac{83}{100}$, 17. $\frac{12}{25}$, 18. $2\frac{1}{2}$, 19. $\frac{27}{50}$, 20. $\frac{3}{40}$, 21. $\frac{9}{50}$, 22. $6\frac{1}{25}$, 23. $\frac{27}{40}$, 24. $3\frac{1}{20}$

Page 54 1. $<$, 2. $>$, 3. $>$, 4. $<$, 5. =, 6. $>$, 7. $<$, 8. $<$, 9. $>$, 10. $>$, 11. $<$ 12. =, 13. =, 14. $>$, 15. $>$ 16. $\frac{2}{11}, \frac{2}{9}, \frac{3}{5}, \frac{3}{4}$ 17. $\frac{2}{9}, \frac{2}{7}, \frac{2}{5}, \frac{2}{3}$ 18. $\frac{1}{4}, \frac{2}{5}, \frac{3}{7}, \frac{1}{2}$ 19. $\frac{5}{8}, \frac{2}{3}, \frac{13}{18}, \frac{8}{9}$, 20. $\frac{4}{5}, \frac{5}{6}, \frac{6}{7}, \frac{7}{8}$, 21. $\frac{1}{5}, \frac{1}{4}, \frac{1}{3}$

Page 55 1. $\frac{5}{9}$, 2. $\frac{3}{7}$, 3. $\frac{9}{10}$, 4. $\frac{4}{5}$, 5. $\frac{8}{9}$, 6. $\frac{2}{3}$, 7. $\frac{1}{2}$, 8. $\frac{1}{3}$, 9. $\frac{3}{17}$, 10. $\frac{3}{8}$, 11. $\frac{5}{6}$, 12. $\frac{27}{28}$, 13. $\frac{3}{5}$, 14. $\frac{2}{5}$, 15. $\frac{2}{15}$, 16. $\frac{71}{82}$, 17. $\frac{1}{4}$, 18. $\frac{1}{3}$, 19. $\frac{21}{32}$, 20. $\frac{3}{4}$, 21. $\frac{37}{100}$, 22. $\frac{2}{3}$, 23. $\frac{2}{3}$, 24. $\frac{4}{5}$, 25. $\frac{2}{4} = \frac{8}{16}$, $\frac{2}{8} = \frac{4}{16}$, $\frac{4}{2} = \frac{16}{8}$, $\frac{8}{2} = \frac{16}{4}$

Page 56 1. $1\frac{2}{7}$, 2. $1\frac{3}{4}$, 3. $2\frac{2}{5}$, 4. 1, 5. $2\frac{1}{2}$, 6. $3\frac{2}{5}$, 7. 3, 8. $2\frac{1}{3}$, 9. $4\frac{1}{3}$, 10. $3\frac{2}{7}$, 11. $1\frac{1}{3}$, 12. $5\frac{1}{4}$ 13. $4\frac{2}{5}$, 14. $6\frac{1}{3}$, 15. $1\frac{1}{5}$, 16. $\frac{19}{5}$, 17. $\frac{8}{3}$, 18. $\frac{19}{8}$ 19. $\frac{28}{5}$, 20. $\frac{31}{8}$, 21. $\frac{51}{8}$, 22. $\frac{23}{3}$, 23. $\frac{47}{5}$, 24. $\frac{33}{5}$ 25. $\frac{60}{7}$, 26. $\frac{27}{10}$, 27. $\frac{41}{12}$, 28. $\frac{44}{7}$, 29. $\frac{31}{9}$, 30. $\frac{27}{7}$

Page 57 1. because 15 is not prime, 2. $\frac{8}{15}$, 3. a. $\frac{4}{5}$, b. $\frac{1}{5}$, 4. a. $\frac{5}{12}$, b. 100, 5. no; once GCF of numerator and denominator is 1, it is in simplest form. 6. $3\frac{1}{2}$ pies, 7. 2, 8. Ellen

Page 58 1. $\frac{37}{56}$, 2. $\frac{23}{30}$, 3. $\frac{13}{144}$, 4. $\frac{5}{6}$, 5. $\frac{1}{8}$, 6. $\frac{1}{3}$, 7. $\frac{15}{16}$, 8. $\frac{4}{9}$, 9. $\frac{7}{40}$, 10. $1\frac{1}{6}$, 11. $1\frac{8}{35}$, 12. $\frac{1}{8}$, 13. $1\frac{1}{14}$, 14. $\frac{35}{72}$, 15. $1\frac{16}{45}$, 16. $\frac{11}{15}$

Page 59 1. $7\frac{55}{56}$, 2. $1\frac{14}{15}$, 3. $11\frac{3}{10}$, 4. $\frac{14}{15}$, 5. $7\frac{3}{16}$, 6. $5\frac{5}{8}$, 7. $10\frac{13}{16}$, 8. $\frac{5}{8}$, 9. $12\frac{5}{12}$, 10. $2\frac{11}{20}$, 11. $5\frac{5}{6}$, 12. $3\frac{1}{18}$, 13. $8\frac{7}{8}$, 14. $4\frac{3}{10}$, 15. $26\frac{3}{20}$, 16. 6, 17. $4\frac{7}{10}$, 18. $\frac{9}{10}$

Page 60 1. $\frac{1}{9}$, 2. $\frac{1}{12}$, 3. $\frac{1}{5}$, 4. $12\frac{1}{4}$, 5. 1, 6. 18 7. $1\frac{2}{3}$, 8. $15\frac{3}{4}$, 9. $\frac{5}{14}$, 10. 33, 11. $\frac{1}{2}$, 12. $2\frac{9}{20}$, 13. $\frac{2}{7}$, 14. $7\frac{1}{3}$, 15. 6, 16. $6\frac{3}{7}$, 17. $\frac{1}{9}$, 18. 1, 19. $\frac{3}{35}$, 20. $\frac{1}{4}$, 21. $\frac{2}{3}$; $\frac{2}{5}$

Page 61 1. 2, 2. $1\frac{1}{2}$, 3. 10, 4. $1\frac{3}{5}$, 5. 8, 6. $2\frac{2}{9}$, 7. $1\frac{1}{4}$, 8. $\frac{8}{17}$, 9. $\frac{2}{3}$, 10. $2\frac{1}{7}$, 11. $2\frac{1}{23}$, 12. $3\frac{2}{3}$, 13. $4\frac{3}{5}$, 14. $1\frac{1}{9}$, 15. $1\frac{1}{3}$, 16. $6\frac{2}{5}$, 17. 14, 18. $1\frac{7}{16}$

Page 62 1. $1\frac{1}{9}$, 2. $\frac{5}{36}$, 3. $\frac{37}{40}$, 4. $6\frac{1}{4}$, 5. $\frac{2}{9}$, 6. $\frac{1}{20}$ 7. $\frac{11}{15}$, 8. $\frac{1}{5}$, 9. $\frac{18}{25}$, 10. $\frac{11}{162}$, 11. $27\frac{3}{5}$, 12. $\frac{1}{35}$ 13. $\frac{1}{20}$, 14. $\frac{7}{10}$, 15. $\frac{1}{10}$

Page 63 1. $\frac{17}{40}$, 2. $3\frac{3}{8}$, 3. $7\frac{7}{8}$, 4. $9\frac{3}{8}$, 5. \$5.25, 6. 80

Page 64 1. $23\frac{1}{3}$, 2. 8, 3. 5, 4. 6,000, 5. $26\frac{2}{3}$, 6. 15,840, 7. $2\frac{1}{2}$, 8. 32, 9. 24, 10. 252, 11. 8, 12. 24, 13. $2\frac{1}{2}$, 14. 4, 15. 96, 16. 9, 17. $3\frac{1}{2}$, 18. 8, 19. 9,000, 20. 80, 21. $\frac{1}{2}$, 22. a. Harry's, b. 0.2 quarts

Page 65 1. 580, 2. .567, 3. 8,100, 4. 823, 5. .132, 6. 6,900, 7. .329, 8. 62, 9. 4.26, 10. 501, 11. 4.285, 12. 1,250, 13. 72,800, 14. 4,300, 15. 320,000, 16. 2.895, 17. .586, 18. .000165, 19. 3.3 cm, 52.35 mm, 0.48 m, 0.0021 km

Page 66 1. 14 m; $12\frac{1}{4}$ m^2, 2. 57m; 170cm^2 3. 169 cm; 1200 cm^2 4. 48.8 cm; 148.84 cm^2 5. 38 m; 85 m^2 6. 73 yd; 195 yd^2 7. 10.4 ft; 5.95 ft^2 8. 8 ft; 4 ft^2 9. Sample Answer: a 3 x 30 rectangle

Page 67 1. 16, 2. 36, 3. 4, 4. 49, 5. 1, 6. 64, 7. 144, 8. 121, 9. 100, 10. 81, 11. 25, 12. 196, 13. 225, 14. 400, 15. 169, 16. 7, 17. 4, 18. 9, 19. 6, 20. 2, 21. 10, 22. 8, 23. 11, 24. 3, 25. 12, 26. 20, 27. 25, 28. 16, 29. 21, 30. 15, 31. Sample Answer: 64

Page 68 1. 11.4 m, 2. 10 cm, 3. 25 in, 4. 5 ft, 5. 25.1 m, 6. 5.4 in, 7. 7.6 ft, 8. 41 cm, 9. 16.1 ft, 10. 15.8 mm, 11. 20.2 in, 12. 17 ft, 13. yes, 14. no, 15. no, 16. yes, 17. yes, 18. no, 19. yes, 20. no, 21. no

Page 69 1. 45 m^2, 2. 66.6 ft^2, 3. $2\frac{5}{36}$ cm^2, 4. $27\frac{1}{12}$ cm^2, 5. 238 in^2, 6. 219.42 m^2, 7. 150 m^2, 8. $127\frac{1}{6}$ ft^2

Page 70 1. 75.36 m; 452.16 m^2 2. 18.84 ft; 28.26 ft^2 3. 47.1 m; 176.63 m^2 4. 56.52 in; 254.34 in^2 5. 10.99 yd; 9.62 yd^2 6. 188.4 m, 2,826 m^2 7. 22.61 yd; 40.69 yd^2 8. 43.96 in; 153.86 in^2 9. $\approx$ 793.8 in^2

Page 71 1. 20 laps, 2. yes, 3. $6\frac{1}{3}$ yd, 4. $22\frac{1}{2}$ in^2, 5. 3 miles, 6. 1,360 ft, 7. 13.1 m, 8. 452.16 mm^2

Page 72 1. $\frac{5}{9}$, 2. $\frac{3}{4}$, 3. $\frac{2}{7}$, 4. $\frac{5}{1}$, 5. $\frac{1}{3}$, 6. $\frac{9}{32}$,

Page 72 continued 7. $\frac{16}{7}$, 8. $\frac{7}{22}$, 9. $\frac{1}{3}$, 10. no, 11. yes, 12. no, 13. yes, 14. yes, 15. no, 16. no, 17. yes, 18. no, 19. yes, 20. yes, 21. no

Page 73 1. $.81 for 1 pound, 2. 44 miles in 1 hour, 3. 3 cups for 1 pound, 4. $1.29 for 1 disk, 5. $90 for 1 day, 6. 3 people in 1 car, 7. 3 pounds in 1 week, 8. 26 tickets in 1 day, 9. 5 people in 1 row 10. 60 miles in 1 day, 11. $7.25 for 1 tape, 12. 500 tickets in 1 day, 13. $\frac{1}{2}$cup for 1 pound, 14. $60 for 1 day, 15. 45 miles in 1 hour, 16. $6.00 for 1 tape, 17. a. corner market: $.00763 for 1 napkin and $.035 for 1 cup; Ed's $.00745 for 1 napkin and $.0356 for 1 cup, b. napkins: Ed's, cups: Corner Market

Page 74 1. 2, 2. 15, 3. 15, 4. 22.5, 5. 4.8, 6. 16, 7. 12, 8. 2, 9. 6, 10. 2.5, 11. 3, 12. 40, 13. 3, 14. 42, 15. 20, 16. 56, 17. 84, 18. 16, 19. 1, 20. 30, 21. 60, 22. 63, 23. 36, 24. 2.5, 25. $\frac{8}{12} = \frac{12}{18}$

Page 75 1. a. $\frac{6}{8}$, $\frac{9}{12}$, $\frac{30}{40}$, b. $\frac{6}{8} = \frac{3}{4}$, $\frac{9}{12} = \frac{3}{4}$, $\frac{30}{40} = \frac{3}{4}$, 2. Josh's, 3. 39 tickets per day, 4. .646 liters per kilometer, 5. $.15, 6. 48 teachers & 6 admin.

Page 76 1. 91%, 2. 23%, 3. 17%, 4. 28%, 5. 47%, 6. 18%, 7. 15%, 8. 63%, 9. 44%, 10. 33%, 11. 21%, 12. 25%, 13. $\frac{7}{10}$, 14. $\frac{19}{20}$, 15. $\frac{9}{20}$, 16. $\frac{43}{50}$, 17. $\frac{33}{100}$, 18. $\frac{21}{25}$, 19. $\frac{1}{2}$, 20. $\frac{1}{5}$, 21. $\frac{3}{20}$, 22. $\frac{3}{4}$, 23. $\frac{1}{4}$, 24. $\frac{1}{10}$

25.

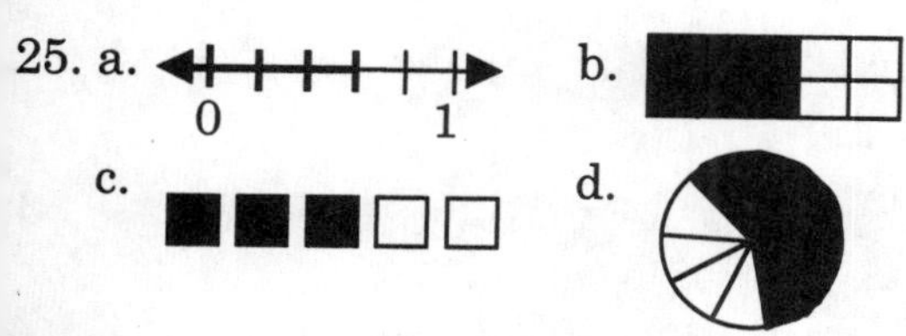

Page 77 1. 38%, 2. 96%, 3. 72%, 4. 40%, 5. 30%, 6. 55%, 7. 67%, 8. 100%, 9. 32%, 10. 83%, 11. 25%, 12. 13%, 13. $\frac{3}{20}$, 14. $\frac{1}{4}$, 15. $\frac{18}{25}$, 16. $\frac{1}{10}$, 17. $\frac{7}{10}$. 18. $\frac{1}{2}$, 19. $\frac{17}{50}$, 20. $\frac{1}{50}$, 21. $\frac{21}{50}$, 22. $\frac{1}{30}$, 23. $\frac{1}{100}$, 24. $\frac{23}{100}$

Page 78 1. 39%, 2. 75%, 3. 87.5%, 4. 10%, 5. 60%, 6. 6%, 7. 4.5%, 8. 99.9%, 9. 46%, 10. 5%, 11. 56.5%, 12. 50%, 13. .43, 14. .89, 15. .07, 16. .02, 17. .062, 18. .345, 19. .9, 20. .17, 21. 1, 22. .51 23. 3.1, 24. .79

Page 79 1. .00032, 2. 3.4, 3. .00014, 4. .000035, 5. 1, 6. .00078, 7. 4.25, 8. .002, 9. 500%, 10. 7,000%, 11. 226%, 12. 3%, 13. 325%, 14. 950%, 15. 590%, 16. 740%, 17. 38,500%, 18. 270%, 19. .35%, 20. 400%, 21. 475%, 22. 820%, 23. 1,050%, 24. 1,225% 25. a. =, b. >, c. <

Page 80 1. $62\frac{1}{2}$%, 2. 24, 3. 64, 4. 43.2, 5. 5, 6. 72.4, 7. 50%, 8. 9, 9. 110.4, 10. 32, 11. 92%, 12. 2, 13. 4, 14. 28%

Page 81 1. 75%, 2. 50%, 3. 20%, 4. 100%, 5. 20%, 6. 21%, 7. 10%, 8. 37%, 9. 100%, 10. 19%, 11. 45%, 12. 29%, 13. 19%, 14. 68%, 15. 2%, 16. no; $14 + discount = original price, but the discount is 30% of the original price, not of $14.

Page 82 1. $3.75, 2. $7.20, 3. $3.43, 4. $3.20, 5. $10.56, 6. $7.50, 7. $.80, 8. $1.40, 9. $13.17, 10. $.38, 11. a. yes; discount comes first because you cannot discount sales tax. b. no; the final price is the same whether you take the discount first and then add the sales tax, or add the sales tax first and then take the discount; final price = $34.24

Page 83 1. $360, 2. 39, 3. 21%, 4. 80%, 5. 48, 6. 150%, 7. a. $70, b. $105

Page 84 1. 35, 42, 49; 2. 256, 1,024, 4,096; 3. 80, 160, 320; 4. 75, 90, 105; 5. $\frac{1}{27}$, $\frac{1}{81}$, $\frac{1}{243}$; 6. 16, 64, 256; 7. 5.4, 6.5, 7.6; 8. 68, 85, 102; 9. .125, .0625, .03125; 10. 30, 36, 42; 11. 625, 3,125, 15,625; 12. 3.2, 6.4, 12.8; 13. 64, 80, 96; 14. 4, 2, 1; 15. a. 10, 10.6, 11.2, 11.8, 12.4; b. 4, 2, 1, $\frac{1}{2}$, $\frac{1}{4}$.

Page 85 1. 5, 10, 20, 40, 80, 160, 320; no; by the 7th day she's exercising 5 1/3 hours. 2. 10 hours, 3. 96 cookies, 4. 2,048 bottles

Page 86

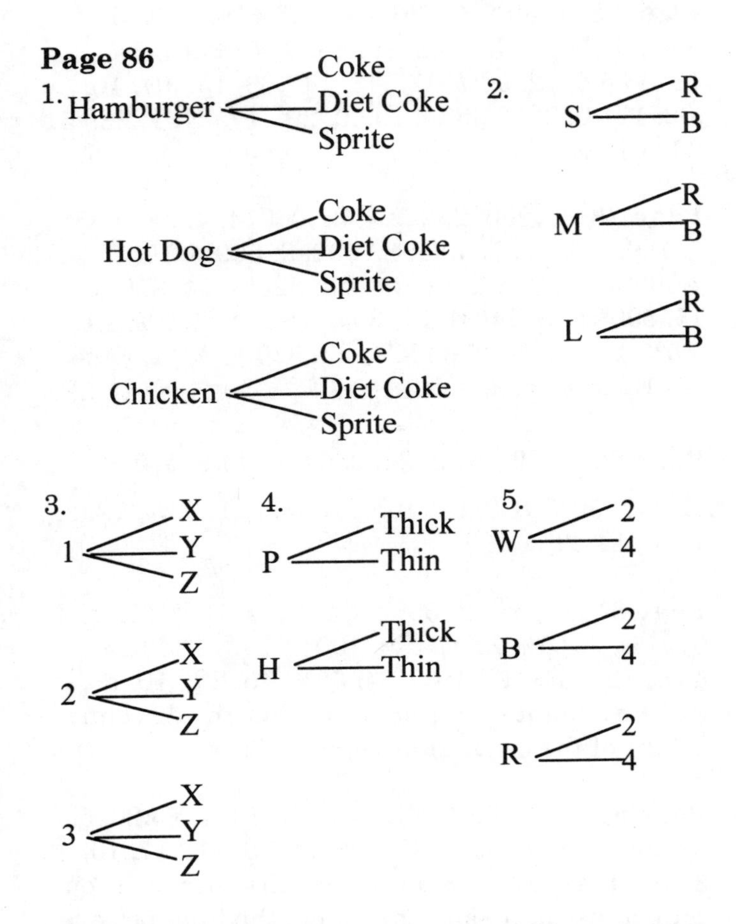

Page 87 1. 168, 2. 40, 3. 90, 4. 48, 5. 16, 6. 120, 7. 36

Page 88 1. $\frac{2}{3}$, 2. a. $\frac{3}{5}$, b. The theoretical probability is $\frac{1}{2}$, which is a little lower than the experimental probability.
3. a. $\frac{1}{2}$, b $\frac{1}{4}$, c. $\frac{1}{4}$, d. $\frac{1}{2}$, 4. a. $\frac{4}{7}$, b. $\frac{5}{7}$

Page 89 1. 729, 2. 96, 3. 255, 4. a. $\frac{13}{60}$, b. $\frac{31}{60}$, c. $\frac{17}{20}$, d. $\frac{7}{60}$, 5. 24